Versatile Hardware Analysis Techniques

Lucas Klemmer • Daniel Große

Versatile Hardware Analysis Techniques

From Waveform-based Analysis to Formal Verification

Springer

Lucas Klemmer
Institute for Complex Systems
Johannes Kepler University Linz
Linz, Austria

Daniel Große
Institute for Complex Systems
Johannes Kepler University Linz
Linz, Austria

ISBN 978-3-031-83092-1 ISBN 978-3-031-83093-8 (eBook)
https://doi.org/10.1007/978-3-031-83093-8

This Springer imprint is published by the registered company Springer Nature Switzerland AG
The registered company address is: Gewerbestrasse 11, 6330 Cham, Switzerland

To Manfred and Alia

Preface

With the slowing-down of Moore's law, the industry is scrambling to keep up with the still growing demand for powerful hardware designs. One strategy to cope with these demands is the move to increasingly application-specific designs, of which AI accelerators are a prime example. Another example of this move to application-specific designs is the emergence of the RISC-V Instruction Set Architecture (ISA). It is not only royalty-free but also highly modular, thus leading to an extensive offering of RISC-V processors covering all application domains from embedded to high-performance computing. Naturally, design complexities and design sizes are still growing rapidly, thus taken together, creating many new challenges for Electronic Design Automation (EDA) toolchains, especially for the labor-intensive verification and debugging phases.

This book proposes several versatile hardware analysis techniques that tackle existing and new challenges. These techniques cover different phases of the hardware development process, including the verification, debugging, and post-synthesis optimization phases. The first contribution is a novel processor verification methodology based on a new application of equivalence checking. The second contribution is a framework for formally verifying microcode. The third contribution is the Waveform Analysis Language (WAL), a Domain Specific Language (DSL) for analyzing waveforms. We evaluate WAL in multiple case studies, including analyzing processor performance metrics for various RISC-V processors and the integration of WAL into a Hardware Description Language (HDL) in the form of analysis passes. The fourth and final contribution is a netlist optimization methodology based on external constraints. All contributions and their effectiveness are presented in detail and are evaluated in extensive experiments and case studies. Furthermore, all contributions are fully implemented and made available as open-source software.

Linz, Austria

Lucas Klemmer
Daniel Große

Acknowledgments

We would like to particularly appreciate all those who have contributed to the results included in this book. Our special thanks go to our collaborators, including Oscar Gustafsson and Frans Skarman from the Division of Electronics and Computer Engineering (ELDA) at Linköping University and Eyck Jentzsch from MINRES Technologies GmbH. We would like to express our appreciation to all the colleagues at the Institute for Complex Systems at the Johannes Kepler University Linz. Finally, we want to thank our families for their unlimited support and love.

Linz, Austria
November 2024

Lucas Klemmer
Daniel Große

Contents

Acronyms

ABI	Application Binary Interface
AI	Artificial Intelligence
APB	Advanced Peripheral Bus
ASIC	Application Specific Integrated Circuit
AST	Abstract Syntax Tree
BMC	Bounded Model Checking
BNF	Backus-Naur Form
BRAM	Block RAM
CFG	Control Flow Graph
CNF	Conjunctive Normal Form
CPI	Cycles per Instruction
CPU	Central Processing Unit
CSR	Control and Status Registers
DCs	Don't Cares
DPLL	Davis-Putnam-Logemann-Loveland
DSL	Domain Specific Language
DUV	Design Under Verification
EC	Equivalence Checking
EDA	Electronic Design Automation
EPEX	Equivalent Program EXecution
ESL	Electronic System Level
FPGA	Field-Programmable Gate Array
FSM	Finite State Machine
FSYN	Formal SYNthesis
HDL	Hardware Description Language
HIR	High-level Intermediate Representation
HPC	High Performance Computing
HW	Hardware
IP	Intellectual Property
IPC	Instructions per Cycle
ISA	Instruction Set Architecture

ISS	Instruction Set Simulator
MIR	Mid-level Intermediate Representation
ODCs	Observability Don't Cares
OISC	One Instruction Set Computer
PC	Property Checking
PDAT	Property-Driven Automatic Transformation
PSL	Property Specification Language
QED	Quick Error Detection
RCA	Ripple-Carry Adder
REPL	Read-Eval-Print-Loop
RISC	Reduced Instruction Set Computer
RTL	Register Transfer Level
SAT	Boolean Satisfiability Problem
SDCs	Satisfiability Don't Cares
SMT	Satisfiability Modulo Theories
SoC	System on a Chip
SQED	Symbolic Quick Error Detection
SVA	SystemVerilog Assertions
SW	Software
TLM	Transaction Level Modeling
UVM	Universal Verification Methodology
VCD	Value Change Dump
VLSI	Very Large-Scale Integration
VP	Virtual Prototype
WAL	Waveform Analysis Language
WAWK	Waveform AWK
XDCs	External Don't Cares

Chapter 1
Introduction

With the slowing down of Moore's law, the times of ever-increasing processor performance gains through optimizations in the manufacturing process come to an end. However, expectations for improved performance and efficiency have not slowed down. Instead, developments such as Apples M1 processor or the unprecedented need for powerful processing units for Artificial Intelligence (AI) tasks nourish the expectations from customers and businesses for ever-increasing performance.

In an attempt to keep up with these rising expectations, large parts of the industry are pivoting toward highly application-specific designs [7]. These highly domain-specific designs are built for a single application only or for a small group of similar applications. One prominent example of this is the intense competition for AI accelerators. Today, commercial accelerators are available from many companies in the digital design and EDA space [3, 28, 29]. The same development can be observed in academia too, where recently, AI-related research became one of the hot topics. Another prime example is the emergence and fast adoption of the modular and open-standard RISC-V ISA [30]. By separating most parts of the ISA into optional extensions, RISC-V scales from the tiniest embedded systems up to the largest High Performance Computing (HPC) applications. Further, RISC-V not only allows fine-grained control over which extension is used, but it also allows users to define their own custom instructions. The customizability and the openness of RISC-V have led to the availability of a wide range of Intellectual Property (IP) and increased industry adoption.

Of course, a drive to more application-specific designs has meant no slowdown in the ever-increasing size and complexity of designs. Taken together, this means that hardware development is not only getting more complicated due to increasing complexity but also due to a significantly increasing number of designs. From the design perspective, much work has been done recently to accommodate these challenges. This includes work on modern code-generator-based Hardware Description Languages (HDLs) [1, 2, 24, 27] that allows extensive parameterization to enable

L. Klemmer, D. Große, *Versatile Hardware Analysis Techniques*,
https://doi.org/10.1007/978-3-031-83093-8_1

quickly adapting to varying needs or the continued work on Virtual Prototypes (VPs) that help in design space exploration and allow earlier software development.

However, while architecting and designing next-generation systems is certainly a daunting task, about 50% of the development process is not spent in the design stage, but during the verification and debugging phases [5]. Both, verification and debugging, can be understood as *hardware analysis problems*, since they try to uncover problems either by analyzing simulations or by statically analyzing the behavior of the design during formal verification. They typically require large amounts of manual work (e.g., writing properties or debugging inside a waveform viewer), and it is therefore imperative to streamline them as much as possible. Considering the huge work requirements, the rapidly increasing number of highly application specific and complex designs, and the low supply of domain experts, it is clear that manual work will not be able to cover all hardware analysis needs. Therefore, a push to automate more analysis tasks is required.

During development of a new hardware design, multiple phases are passed. At a very high level, the hardware development process can be broken down into seven phases: *specification*, *design*, *verification*, *debug*, *synthesis*, *post synthesis optimization*, and *fabrication*. These phases and their relation to each other are shown in Fig. 1.1. After the *specification* is ready, the *design* space of possible implementations is typically explored using high-level models such as VPs at the Electronic System Level (ESL) [4, 9] leveraging Transaction Level Modeling (TLM). If the high-level architecture is decided upon, it is commonly refined to the hardware level using an Register Transfer Level (RTL) description.

Then, the RTL model is *verified*. Here, two kinds of verification can be distinguished: *Functional* verification checks the error-free behavior of the implementation, i.e., its role is to analyze a design, looking for functional bugs that violate the specification. In contrast, *extra-functional* verification covers an extensive range of analysis tasks [8]. These can include the computation of average latencies, the interaction of hardware and software, or the generation of complimentary data such as charts or reports from simulations. If the verification spots an extra-functional problem, for example, a critically high utilization of a bus during runtime, that can become a problem later on, significant changes to the design might be required. However, in the end, both functional and extra-functional verification are consulted to determine if an implementation is specification-compliant.

The main workhorse for hardware verification is simulation. During hardware simulation, stimuli are applied to a Design Under Verification (DUV) to drive the design for a given amount of time. Manually writing stimuli that ensure hard-to-reach corner cases are covered is a nontrivial task. As a consequence, automated stimuli generation has been an actively researched area [22, 23, 31].

After the simulation, the correctness and the performance of the DUV have to be evaluated. This is commonly done by writing properties that specify the correct behavior of the system against which the simulation is compared. If such a property is violated, the verification engineer typically researches the root cause of the bug by opening a simulation dump (also called a waveform) inside a *waveform viewer*. Waveform viewers are software tools which allow viewing signal values over the

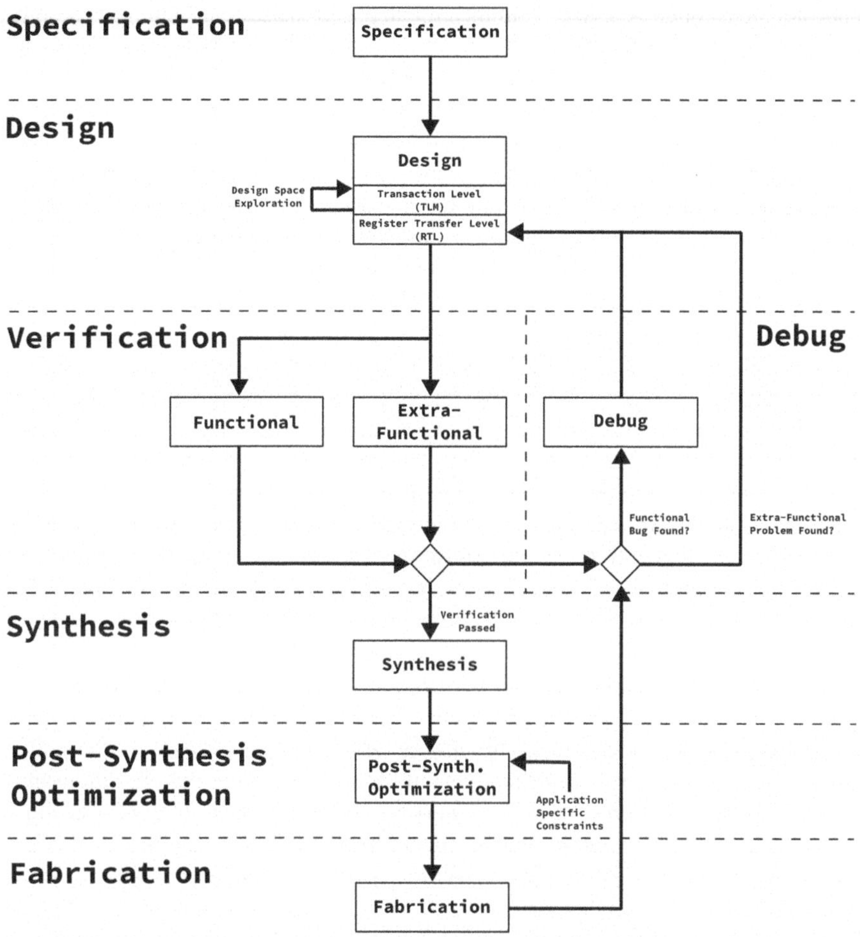

Fig. 1.1 A high-level view of the hardware development process

simulation time. Modern waveform viewers provide many tools that help engineers in analyzing a waveform, still, while all these features help in understanding and debugging, *waveform viewing is a highly manual and tedious process*. Since waveforms contain highly detailed information about the micro-architectural state during the simulation, they provide a rich information source about many aspects of a design. Information such as bus performance, cache utilization, and even power estimates can be derived by inspecting waveforms. However, due to the large amount of data that has to be considered, manual analysis of such information is infeasible. While commercial EDA suites provide many advanced tools to analyze hardware simulations, they can be used exclusively for use-cases which have been considered by the vendor. Further, commercial tools tend to be highly monolithic and locked down, which makes combining and building new workflows challenging. On the other side in the open-source community, much progress has been achieved recently. Regardless of this, especially in terms of verification, the open-source toolchains are still trailing behind their commercial counterparts.

If a bug is uncovered during the verification phase, it has to be localized and fixed in the following debug phase. This phase typically involves a significant amount of manual work, often spent inside waveform viewers. In addition, debugging is typically an iterative task, where a solution is narrowed down over multiple debug-verification cycles. The consequence of this is that much time is spent during the debugging phase. Hence, improving the debugging speed and reducing the number of required debug-verification iterations could lead to significant reductions of the hardware development time.

After a design has passed all the previous phases, it is *synthesized* to a netlist. During synthesis, various kinds of logic optimizations are performed to get a result satisfying the performance, timing, or energy requirements. Now, the netlist is either programmed onto a Field-Programmable Gate Array (FPGA) or pushed all the way down to being taped out as an Application Specific Integrated Circuit (ASIC). Still, a final phase before this process can be identified the *post-synthesis* optimization. By considering application-specific constraints, such as a guaranteed absence of certain input combinations, an otherwise final netlist could be further optimized. This is an interesting scenario, since optimizations can be performed even without direct access to the design sources and without running the full synthesis steps again. However, this kind of netlist optimization is challenging due to the low-level nature of the format. Elements which have been high-level structures before turn into individual gates and wires. This makes larger manual changes to the netlist basically impossible, since any change would require rewriting possibly thousands of gates. To solve this problem, initial work has been presented, though it is either not available or its usefulness and practicability is limited. Examples of this are tools which are based completely on proprietary tools, which hinders free research, follow-up studies, or replications, as well as research tools that are focused on theoretic or academic settings [26]. In conclusion, the post-synthesis optimization of netlists is a further phase in the hardware design process that can provide additional benefits even in situations where no design source is available. Yet, this kind of optimization has only been considered recently and thus leaves much room for improvements and interesting work.

Ultimately, once all phases have been successfully completed, the design can be manufactured and has to be then verified using post-silicon techniques. According to [5] though, more than three quarters of all ASIC designs require more than two respins until the fabricated design satisfies all requirements. At this late stage in the development process, bug fixing is significantly more costly than during the pre-silicon stages. Therefore, eliminating as many problems as possible before fabrication is critical, and improvements in the earlier phases are essential to keep delivering next-generation systems that satisfy all expectations.

1.1 Book Contribution

As presented, the hardware development process is facing significant challenges with the growing expectations on the performance and the increasing number of highly application-specific designs. To counteract these problems, this book presents *versatile and automated hardware analysis techniques*. These techniques cover and extend most parts of the hardware design process and are depicted in Fig. 1.2; there our proposed techniques are shown in light yellow color. They enhance three of the seven hardware development phases presented above:

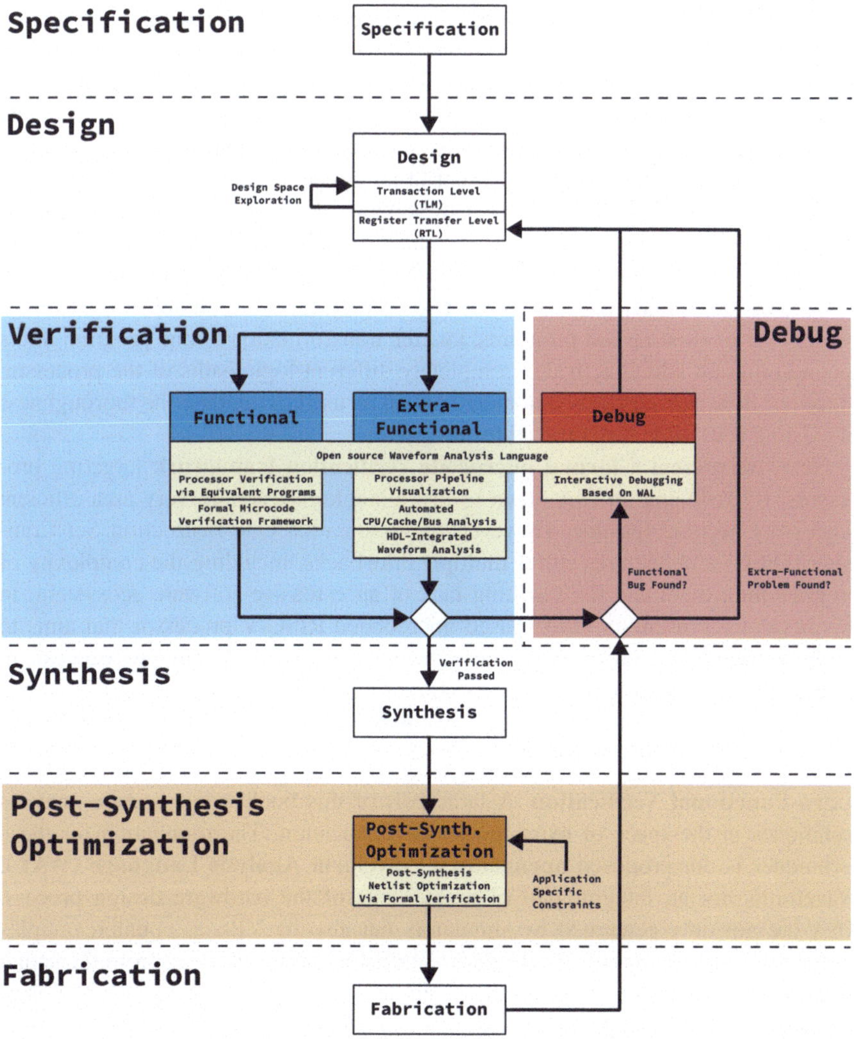

Fig. 1.2 The high-level hardware development process extended by the versatile and automated hardware analysis techniques presented in this book

1. (Extra-)Functional verification (blue)
2. Debug (red)
3. Post-synthesis optimization (orange)

In all of these phases, we present innovative approaches which have been fully implemented and evaluated in several experiments and extensive case studies. In particular, the majority of our experiments leverage open-standard and open-source technologies such as RISC-V to ensure easy access and to allow reproducibility. In the following, we will walk through the phases covered in this book, describing each proposed approach (light yellow boxes in Fig. 1.2).

Functional Verification In this category, a novel Equivalence Checking (EC)-based approach for **processor verification via equivalent programs** is presented in this book. By utilizing a formal model of the processor, we generate two distinct programs which are proven equivalent and thus should result in exactly the same architectural state. Now, both programs are run on two instances of the same processor, and the resulting architectural states of both instances are compared. If they don't match, a bug must be present in the processor design. This verification strategy leverages formal verification to generate the required simulation stimuli (the instructions making up the two generated programs). It also removes the need to write properties that check the correct behavior of the processor, since this is already covered by the equality property of the two programs. An additional use-case is the *broadening* of existing test programs, i.e., the transformation of one program into a second program which activates completely different logic paths of the processor. Since our flow is completely automated, it can be used to improve the thoroughness of existing tests with very little additional effort.

Next, we present a **formal microcode verification framework** targeting processors. By reducing an ISA down to just a single instruction, very area-efficient processors become feasible. However, these so-called One Instruction Set Computer (OISC) architectures suffer multiple drawbacks, including the complexity of programming them and the resulting lack of an extensive software ecosystem. In this book, we present an OISC-based microcoded RISC-V processor that aims to combine the advantages of OISCs with the extensive RISC-V software ecosystem. To ease the development of the microcode, we created a VP to allow fast iterations and design space exploration. Finally, we employ our formal verification framework to the developed OISC microcode.

Extra-Functional Verification A large part of this book is focused on analysis techniques in the space of extra-functional verification. The foundation for these techniques is our proposed **open-source Waveform Analysis Language (WAL)**. Waveforms are an integral part of many steps of the hardware design process. They are not only generated by simulators but also by VPs, as counterexamples by formal verification tools, or they are sampled by logic analyzers from the actual hardware. Waveform analysis has the potential to be employed at various phases of the hardware development process. Nonetheless, today waveform analysis is limited mostly to manually inspecting them inside a waveform viewer or to the analysis functions included in large monolithic and proprietary EDA tools.

With WAL, we present the first programming language targeted toward automating analysis problems in the hardware development process. By being designed as a DSL and not as an ad hoc extension to a common software programming language, WAL provides language features specifically designed for waveform analysis using a concise and natural syntax. The core idea behind WAL is that everything which is integral to hardware design is a natural part of the language. This includes direct access to all signals, the design hierarchy, and, maybe most importantly, the time which is inherent to hardware design. Since WAL is a programming language and not a single-purpose application, it allows writing analysis programs from small one-time scripts up to elaborate and reusable analysis frameworks. In fact, WAL is not only applicable to extra-functional analysis but also during functional verification and debugging. Further, because it is designed as a versatile tool, it can be employed in other roles as well.

WALs generic programming features allow writing reusable analysis programs that are not tied to a specific design. We show how this can be leveraged in several examples including an interactive **processor pipeline visualization** and two extensive case studies. In particular, we show how WAL can be used for **automated CPU/Cache/Bus analysis**. With WAL, central performance metrics such as the Instructions per Cycle (IPC) and the behavior of instruction and data caches can be analyzed in a general and processor independent way. Further, we show how WAL analysis programs can be made accessible to HDL developers by embedding WAL into Spade [24], a novel HDL. By utilizing the Spade HDL compiler's knowledge about design internals, we can connect premade analysis passes to the design without significant work required from the side of the designer. In this manner, hardware designs can be complemented with complex design analysis programs that are ready to be used by users with minimal effort.

Debug If a bug is uncovered during verification, it has to be fixed in the following debugging phase. This typically entails multiple iterations of introducing changes to the design followed by a verification run to validate the changes in a design-debug-verification cycle. Due to often very long-running verification setups (i.e., simulation or formal verification runtime), this can significantly slow down the overall development speed. In addition, once an error is found, the simulation ends, and typically a waveform is produced. However, this also means, that the simulation context in which the bug occurred is left, and the debugging happens in a different context. To reduce the iteration time of this design-debug-verification and to provide a highly integrated debugging environment, we present a novel **interactive debugging flow based on WAL**. This flow is based on the idea of injecting new signals into already existing waveforms. The values of these new virtual signals are driven by WAL programs. Since WAL is a powerful language, virtual signals can be defined that emulate combinational or sequential behavior, and they can contain all WAL values including lists or hash maps. By injecting virtual signals that shadow faulty signals from the waveform, developers can quickly try bug fixes without having to run the full simulation or verification setup. Further, the proposed bug-fix can be directly analyzed and verified inside the WAL environment.

Post-Synthesis Optimization Finally, we present a **post-synthesis netlist optimization via formal verification**. This approach is based on the observation that highly application-specific external constraints can be utilized to effectively remove parts of a netlist without compromising on correctness. For example, if the program running on a small embedded device is not using any division instructions, the gates implementing the division can be safely removed from the netlist. Yet, identifying those gates is a challenging task considering the enormous number of gates in a design and the resulting complexity of the formal problems. Our research in particular leverages an open-source-first approach, considering the required tools. This allows us to exploit the high parallelization potential of netlist optimization by distributing it over multiple machines, since we are free from license limitations. Further, our approach provides multiple back ends that can be used to perform the optimization. These back ends include Property Checking (PC) and EC as well as the possibility to perform initial simulation-based rule filtering using WAL.

As a matter of principle, all of our implementations are made *available as open-source software* to stimulate further research and to make all of our results reproducible. This includes the following list of available resources:

- **EPEX: Equivalent Program Execution**
  Source code: https://github.com/ics-jku/epex-formal-rv32-model
- **Microcoded RISC-V Platform VP**
  Source code: https://github.com/ics-jku/goldcrest-vp
- **Formal Microcode Verification Framework**
  Source code: https://github.com/ics-jku/goldcrest-microcode-verification
- **WAL: Waveform Analysis Language**
  Documentation: https://wal-lang.org
  Source code: https://github.com/ics-jku/wal
- **WAL HDL Integration**
  Source code: https://gitlab.com/spade-lang/wal_analysis
- **WAL SystemVerilog Assertions Compiler**
  Source code: https://gitlab.com/lklemmer/wsva
- **Post Synthesis Netlist Optimization**
  Source code: https://github.com/ics-jku/fsyn

This book is structured as follows: First, Chap. 2 provides relevant background information on RISC-V and formal methods. Afterward, Chap. 3 presents a novel processor verification methodology based on equivalence checking. This chapter is based on work published in [10]. Subsequently, Chap. 4 introduces a framework for the formal verification of a microcoded RISC-V platform. This chapter is based on work published in [11, 18]. Next, Chap. 5 introduces the Waveform Analysis Language (WAL) and several case studies that demonstrate how complex waveform analysis problems can be solved using WAL. This chapter is based on work published in [12, 14, 15, 17, 21]. The next three chapters present applications of

WAL. First, Chap. 6 presents a processor analysis methodology based on WAL which can be used to analyze detailed performance metrics of processors, processor pipelines, and caches. This chapter is based on work published in [13, 19]. Then, Chap. 7 presents how complex hardware analysis tasks can be made available to hardware developers easily by embedding WAL into an Hardware Description Language (HDL), and Chap. 8 presents a new interactive debugging workflow based on WAL. These chapters are based on work published in [25] and in [16]. Subsequently, Chap. 9 presents a post-synthesis netlist optimization framework leveraging external constraints. This chapter is based on work published in [6, 20]. Finally, the book is concluded in Chap. 10.

References

1. C.P. Baaij, Digital circuits in CλaSH. PhD. Thesis, University of Twente (2015)
2. J. Bachrach, H. Vo, B.C. Richards, Y. Lee, A. Waterman, R. Avizienis, J. Wawrzynek, K. Asanovic, Chisel: constructing hardware in a Scala embedded language, in *Design Automation Conference* (2012), pp. 1212–1221
3. Cadence accelerates on-device and edge AI performance and efficiency with new neo NPU IP and neuroweave SDK for silicon design. https://www.cadence.com/en_US/home/company/newsroom/press-releases/pr/2023/cadence-accelerates-on-device-and-edge-ai-performance-and.html. Accessed 19 April 2024
4. T. De Schutter, *Better Software. Faster!: Best Practices in Virtual Prototyping* (Synopsys Press, Sunnyvale, 2014)
5. H.D. Foster, The 2020 Wilson research group functional verification study. https://blogs.sw.siemens.com/verificationhorizons/2021/01/06/part-8-the-2020-wilson-research-group-functional-verification-study
6. D. Große, L. Klemmer, D. Bonora, Using formal verification methods for optimization of circuits under external constraints, in *Design, Automation and Test in Europe* (2024), pp. 1–6
7. J.L. Hennessy, D.A. Patterson, A new golden age for computer architecture. Commun. ACM **62**(2), 48–60 (2019)
8. J.L. Hennessy, D.A. Patterson, Understanding multidimensional verification: where functional meets non-functional. Microprocess. Microsyst. **71**, 102867 (2019)
9. V. Herdt, D. Große, R. Drechsler, *Enhanced Virtual Prototyping: Featuring RISC-V Case Studies* (Springer, Berlin, 2020)
10. L. Klemmer, D. Große, EPEX: processor verification by equivalent program execution, in *ACM Great Lakes Symposium on VLSI* (2021), pp. 33–38
11. L. Klemmer, D. Große, An exploration platform for microcoded RISC-V cores leveraging the one instruction set computer principle, in *IEEE Annual Symposium on VLSI* (2022), pp. 38–43
12. L. Klemmer, D. Große, WAL: a novel waveform analysis language for advanced design understanding and debugging, in *ASP Design Automation Conference* (2022), pp. 358–364
13. L. Klemmer, D. Große, Waveform-based performance analysis of RISC-V processors: late breaking results, in *Design Automation Conference* (2022), pp. 1404–1405
14. L. Klemmer, D. Große, A DSL for visualizing pipelines: a RISC-V case study, in *RISC-V Summit Europe* (2023)
15. L. Klemmer, D. Große, Programming language assisted waveform analysis: a case study on the instruction performance of SERV, in *Workshop on Open-Source Design Automation* (2023)
16. L. Klemmer, D. Große, Towards a highly interactive design-debug-verification cycle, in *ASP Design Automation Conference* (2024), pp. 692–697

17. L. Klemmer, D. Große, WAVING goodbye to manual waveform analysis in HDL design with WAL. IEEE Trans. Comput. Aided Des. Circuits Syst. **43**(10), 3198–3211 (2024)
18. L. Klemmer, S. Gurtner, D. Große, Formal verification of SUBLEQ microcode implementing the RV32I ISA, in *Forum on Specification and Design Languages* (2022), pp. 1–8
19. L. Klemmer, E. Jentzsch, D. Große, Programmable analysis of RISC-V processor simulations using WAL, in *Design and Verification Conference and Exhibition Europe* (2022)
20. L. Klemmer, D. Bonora, D. Große, Large-scale gatelevel optimization leveraging property checking, in *Design and Verification Conference and Exhibition Europe* (2023), pp. 86–93
21. L. Klemmer, F. Skarman, O. Gustafsson, D. Große, Surfer: a waveform viewer as dynamic as RISC-V, in *RISC-V Summit Europe* (2024)
22. Y. Naveh, M. Rimon, I. Jaeger, Y. Katz, M. Vinov, E. Marcu, G. Shurek, Constraint-based random stimuli generation for hardware verification. AI Mag. **28**(3), 13–13 (2007)
23. RISC-V DV. https://github.com/chipsalliance/riscv-dv
24. F. Skarman, O. Gustafsson, Spade: an expression-based HDL with pipelines, in *3rd Workshop on Open-Source Design Automation* (2023)
25. F. Skarman, L. Klemmer, O. Gustafsson, D. Große, Enhancing compiler-driven HDL design with automatic waveform analysis, in *Forum on Specification and Design Languages* (2023), pp. 1–8
26. M. Soeken, H. Riener, W. Haaswijk, E. Testa, B. Schmitt, G. Meuli, F. Mozafari, S.-Y. Lee, A.T. Calvino, D.S. Marakkalage, G.D. Micheli, The EPFL logic synthesis libraries (2022)
27. SpinalHDL contributors. SpinalHDL. https://github.com/SpinalHDL/SpinalHDL (2022)
28. Synopsys introduces industry's highest performance neural processor IP. https://news.synopsys.com/2022-04-19-Synopsys-Introduces-Industrys-Highest-Performance-Neural-Processor-IP. Accessed 19 April 2024
29. Tenstorrent tensix AI IP. https://tenstorrent.com/ip/tensix-ai-ip/. Accessed 19 April 2024
30. A. Waterman, K. Asanović, *The RISC-V Instruction Set Manual; Volume I: Unprivileged ISA*. SiFive Inc. and CS Division, EECS Department, University of California, Berkeley (2019)
31. S. Yang, R. Wille, D. Große, R. Drechsler, Minimal stimuli generation in simulation-based verification, in *EUROMICRO Symposium on Digital System Design* (2013), pp. 439–444

Chapter 2
Background

This chapter introduces the concepts necessary to understand the hardware analysis techniques presented in the rest of this book. It starts with an introduction to the RISC-V ISA, which is used in most chapters of this book. Since formal methods provide the foundation for several approaches of this book, we also give a brief introduction to formal methods commonly used in EDA tools.

2.1 RISC-V

RISC-V is a royalty-free ISA first proposed by researchers from UC Berkeley [28]. Today, it is controlled by RISC-V International, a nonprofit organization. As suggested by the name, RISC-V is a Reduced Instruction Set Computer (RISC) architecture, and the latest member of a family of ISAs developed at UC Berkeley [15, 16, 21, 26]. From the get-go, RISC-V was designed to be royalty free, meaning that anybody can freely implement RISC-V processors without having to pay license fees, and accommodating to the open-source spirit. While RISC-V is not the only royalty-free ISA [14, 24], it is the first to achieve significant traction with industry, researchers, and even with hobbyist communities. Today, RISC-V cores are available from various companies targeting different market segments, ranging from embedded devices all the way to AI inference platforms. Contrary to most established ISAs today and enabled by the permissive license, RISC-V cores are not only available from companies but also from several open-source repositories. These open-source RISC-V cores are often developed and maintained by academic institutions [8, 25] or small companies [22, 27].

This wide range of domains and IP providers is in part due to the modular and extendable architecture of RISC-V. The RISC-V ISA is organized around a base instruction set, and several optional extensions, of which some examples will be presented later in this section. Compared to other ISAs, the RISC-V

L. Klemmer, D. Große, *Versatile Hardware Analysis Techniques*,
https://doi.org/10.1007/978-3-031-83093-8_2

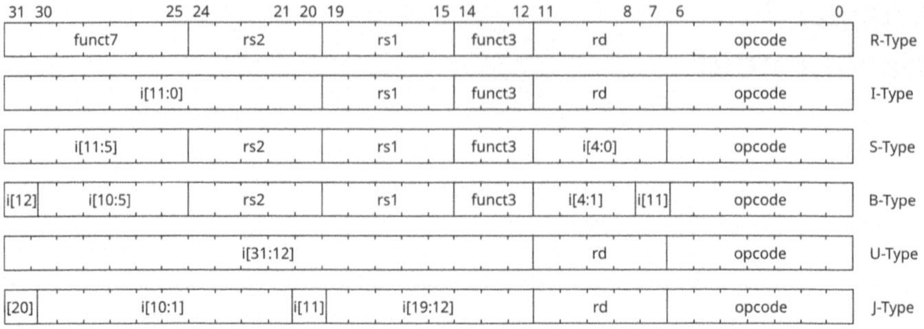

Fig. 2.1 Instruction types of the RV32I base instruction set

base instruction set is limited, containing only basic integer instructions, going as far as omitting multiplication and division. The RISC-V specification also has two variants, supporting 32-bit and 64-bit address ranges and word sizes. In the rest of this book, we will focus only on the 32-bit variant. As a result of this modularity, RISC-V scales from the smallest embedded devices up to high-performance computing applications. Further, this also lowered the entry barrier to developing RISC-V cores significantly, allowing small groups or even individuals to develop specification-compliant designs.

The RV32I base instruction set defines 32, 32-bit wide general purpose registers called `x0`, `x1`, ..., `x31`. Register `x0` is hardwired to the value 0. Each RV32I register also has a second Application Binary Interface (ABI) name that represents the function of this register in the RISC-V calling convention. However, these names only describe the function of the registers without limiting their usage. For a list of these ABI register names, we refer the reader to the "RISC-V Assembly Programmer's Handbook" [28]. Additionally, there is also a user-accessible register holding the current program counter called *pc* and some control and status registers that are accessible through the Control and Status Registers (CSR) instructions. All RV32I instructions use a 32-bit fixed length layout that can be separated into the six instruction formats as shown in Fig. 2.1.

An exemplary program in RISC-V assembly is shown in Listing 2.1. This program sums up the integers in a zero-terminated memory region defined on Line 2. First, the address of the data region is loaded into register `x10` using the `la` pseudo-instruction on Line 6. Then, the result register `x11` is set to 0, and the main loop is entered. There, the next datum is loaded into register `x20` on Line 9. If the loaded datum is zero, the loop is exited. This is checked using the `beq` branching instruction. Next, the datum is added to the result register, and the pointer into the data region is moved to the next entry by adding the number of bytes in a word (4 bytes in RV32I) to it. Finally, the program starts the next iteration of the loop by jumping up to the start of the loop using the `jal` instruction on Line 13. `Jal` stores the address of the instruction following the jump into the result register. Since we are not interested in this link address, we discard it by writing into register `x0`, which is write-protected and always contains the value 0.

```
.data
data: .word 1, 2, 3, 4, 0

.text
main:
    la x10, data        # pointer into data
    li x11, 0           # holds result
loop:
    lw x20, 0(x10)      # load next datum
    beq x20, x0, end    # check if end is reached
    add x11, x11, x20   # add datum to result
    addi x10, x10, 4    # move pointer to next datum
    jal x0, loop        # start next iteration
end:
```

Listing 2.1 Exemplary program in RISC-V assembly

2.2 Formal Methods

Due to exponential scaling, practically every relevant hardware design in use today is virtually impossible to verify by experimentally testing every possible input combination. This problem is further exacerbated when considering that sequential circuits have to be examined over multiple clock cycles. Thus, achieving full coverage of hardware designs by using purely simulation-based verification is practically impossible. In particular, while simulation might be adequate for common cases, many intricate bugs hide in hard-to-find corner cases that are very challenging to hit using simulation. Stepping back from the brute-force approach of simulation, formal methods have proven to be effective at finding bugs, including those hard-to-find corner cases. This capability of formal methods is based on well-studied theoretical foundations, in particular the Boolean Satisfiability Problem (SAT) and Satisfiability Modulo Theories (SMT) theories.

In the rest of this section, we will present a brief application-level introduction to SAT and SMT solving. For an in-depth introduction to SAT and SMT, we refer the reader to [6].

2.2.1 SAT

The Boolean Satisfiability Problem (SAT) problem formulates the question, if there exists a variable assignment that satisfies a given Boolean formula. More formally, given a Boolean formula $f(x_1, x_2, \ldots, x_n)$, is there an assignment to the variables $x_1, x_2, \ldots, x_n$ that will satisfy the formula f. If such an assignment exists, the formula is *satisfiable*, if not, it is *unsatisfiable*. Typically, Boolean formulas are represented in Conjunctive Normal Form (CNF) to facilitate machine processing and automated reasoning. A Boolean formula in CNF form is a conjunction of a

Table 2.1 All possible input combinations for Boolean function g

x_1	x_2	$x_1 \vee x_2$	$x_1 \vee \overline{x}_2$	$\overline{x}_1 \vee x_2$	$\overline{x}_1 \vee \overline{x}_2$	g
0	0	0	1	1	1	0
0	1	1	0	1	1	0
1	0	1	1	0	1	0
1	1	1	1	1	0	0

set of clauses. Each clause consists of literals which are either variables (e.g., x) or negated variables (e.g., $\overline{x}$). If at least one of the literals inside a clause is satisfied, the clause is satisfied too.

$$\bigwedge_{j=1}^{m} \left(\bigvee_{k=1}^{n} x_{j,k} \right)$$

Example 2.1 Consider the following Boolean formula $f(x_1, x_2, x_3) = (x_1 \vee \overline{x}_2) \wedge (x_2 \vee x_3)$. This formula is *satisfiable*, as proven, for example, by the assignment $x_1 = 0, x_2 = 0, x_3 = 1$. In this example, the first clause is satisfied by the negated x_2 literal, which becomes 1 given the assignment $x_2 = 0$. The second clause is satisfied by the assignment $x_3 = 1$.

Example 2.2 Now, consider the following Boolean formula $g(x_1, x_2) = (x_1 \vee x_2) \wedge (x_1 \vee \overline{x}_2) \wedge (\overline{x}_1 \vee x_2) \wedge (\overline{x}_1 \vee \overline{x}_2)$. This formula is *unsatisfiable*, since one of the clauses is false for every variable assignment. Compared to showing that a formula is satisfiable (or *SAT*), where only one satisfying variable assignment has to be found, proving that a formula is unsatisfiable (or *UNSAT*) requires that all possible input combinations evaluate the formula to false. For a low number of variables, this can be achieved using a logic table as shown in Table 2.1.

The SAT problem is one of the fundamental problems in computer science; in fact, it was the first problem to be shown to be $\mathcal{NP}$*-complete* [7]. Due to this, naive techniques fail at solving practically all relevant SAT problems. SAT is also very versatile, as many other problems can be reformulated as a SAT problem. Therefore, much research has been conducted on finding algorithms that can handle practically relevant SAT instances, since improvements to these algorithms directly translate to many other domains as well. Today, most SAT solvers are based on the Davis-Putnam-Logemann-Loveland (DPLL) algorithm [9, 10], which utilizes backtracking and additional optimizations such as unit propagation, efficient decision heuristics, and conflict-driven learning [17] that allow significantly reducing the search space. Even though SAT is $\mathcal{NP}$-complete, many relevant problems can be solved within practical runtimes due to advancements in SAT solving algorithms, heuristics, and very performant SAT solver implementations. Examples for high-performance SAT solvers are Chaff [18], MiniSAT [13], PicoSAT [2], and the parallel solver Gimsatul [4].

Since many problems can be reformulated as SAT problems, SAT solvers form the backbone of many EDA tools. They are commonly used during verification, e.g., for Bounded Model Checking (BMC) [3, 19]. In this scenario, a hardware design and a property that should hold are combined using unrolling and expressed as a Boolean formula. Now, a SAT solver can determine if the resulting function is satisfiable, which allows deriving a counterexample for the temporal property at hand.

2.2.2 SMT

Satisfiability Modulo Theories (SMT) is a generalization of SAT that extends the SAT problem by various additional first-order theories including bit vectors and memories. By extending SAT solvers with these specialized theory solvers, significant performance improvements can be achieved. This is in particular true for problems that make heavy use of some of the specialized theories such as the microcode verification framework which will be presented in Chap. 4. From a high-level perspective, SMT solvers utilize SAT to steer the application of the theory solvers. To achieve this, the SMT solver constructs a SAT problem whose literals represent theory subproblems whose validity is checked by the theory solvers.

Today, many high-performance SMT solvers are available; in fact, regular competitions are held which seek to find the best performing solvers [23]. Prominent examples are Z3 [11], Boolector [20], or Yices [12]. While all of these solvers are very fast, different solvers excel at different tasks. Therefore, selecting the right solver can lead to significant performance improvements. In practice, finding the appropriate solver often requires some experiments. However, since most solvers converged on the single SMT-LIB2 standard [1] for how to encode SMT problems, this is often straightforward.

In the rest of this section, we provide a brief example in which we formulate and solve an SMT problem using an SMT solver.

Example 2.3 In this example, we use an SMT solver to find two terms a and b whose sum s is equal to a given value. In particular, we use bit vector theory to model the operands and the addition operation. By using bit vectors, all values can be limited to a certain number space by specifying bit-widths. The SMT-LIB2 implementation of this problem is shown in Listing 2.2. The SMT-LIB2 format uses a Lisp inspired prefix notation in which all function applications are written in the form of $(f\ a1\ a2\ \ldots\ an)$ where f is the function name or operator and $a1$ to an are the arguments.

First, we enable the quantifier free bit vector theory on Line 1. Then, we configure the solver to produce a model (i.e., output a variable assignment) using the `set-option` function on Line 2. Next, we define three 8-bit constants, two for the terms (a and b) and one for the sum on Lines 4–6. On Line 8, the core operation of this problem, the addition, is implemented. This is done via an `assertion` that constraints the s constant to always hold the value that is the addition of a and b. Please note that the special `bvadd` bit vector addition function is used. In this example, we want to identify two 8-bit numbers whose sum is equal to 14. This is implemented via an additional constraint on the s constant, shown on Line 9. At this point, the s constant is constrained, but both a and b are unconstrained. This allows the solver to freely assign values to both constants. All possible assignments to those two variables represent the search space that is searched by the solver. Finally, the SMT solver is queried on Line 11, and the variable assignment is printed (if the instance is SAT) on Line 12.

```
1  (set-logic qf_bv)
2  (set-option :produce-models true)
3
4  (declare-const a (_ BitVec 8))
5  (declare-const b (_ BitVec 8))
6  (declare-const s (_ BitVec 8))
7
8  (assert (= s (bvadd a b)))
9  (assert (= s (_ bv14 8)))
10
11 (check-sat)
12 (get-model)
```

Listing 2.2 smt-lib2 code modelling a simple addition

Using Z3, the model shown in Listing 2.3 is produced.[1] The constant a is assigned the value 14 (shown in the listing in hexadecimal as $\#x0e$) and b is assigned 0. While this is certainly a valid assignment, it is also "uninteresting" in the sense that one of the constants is already the targeted value. This showcases a common problem when working with SMT solvers. Often, additional guidance is needed to fine-tune the results. For example, when using an SMT solver to synthesize RISC-V instructions, solvers tend to generate instructions without side effects (also called no-ops) that, while valid RISC-V instructions, do not change the architectural state of the processor beside modifying the program counter.

```
(
  (define-fun a () (_ BitVec 8)
    #x0e)
  (define-fun b () (_ BitVec 8)
    #x00)
  (define-fun s () (_ BitVec 8)
    #x0e)
)
```

Listing 2.3 Variable assignment produced by Z3

To solve this problem, additional constraints can be added to the SMT instance, as is shown in Listing 2.4. The two shown assertions constrain both constants a and b to never assume the value 0. Solving the problem again using Z3 results in the model shown in Listing 2.5. Now, a and b have been assigned the values 255 ($\#xff$)

[1] The actual values found by Z3 are not deterministic and might change for further executions.

```
(assert (not (= a (_ bv0 8))))
(assert (not (= b (_ bv0 8))))
```

Listing 2.4 Additional constraints that prohibit the constants to assume the value 0

```
sat
(
  (define-fun a () (_ BitVec 8)
    #x0f)
  (define-fun b () (_ BitVec 8)
    #xff)
  (define-fun res () (_ BitVec 8)
    #x0e)
)
```

Listing 2.5 Variable assignment produced by Z3 after adding additional constraints

and 15 ($\#x0f$). Considering the overflow semantics of bit vector arithmetic, this is also a valid assignment. It also showcases a second property of SMT solvers. They often choose unintuitive assignments which would not have been chosen by humans. This property, however, makes SMT solvers a good fit for many applications as it allows them to find corner-case scenarios.

Finally, if the same solver is used to solve the same problem again, the resulting model will probably be the same. This is a challenge in scenarios in which multiple models should be generated. One solution to this problem is to exclude the already generated models by adding additional constraints to the SMT instance. An example of this is shown in Listing 2.6 where the previously generated model is excluded. After adding this constraint, the solver will generate a different model if such a model exists. For an additional approach to excluding already observed models, see [5].

```
(assert (not
  (and (= a (_ bv255 8))
       (= b (_ bv15 8)))))
```

Listing 2.6 Assertion excluding a possible variable assignment

References

1. C. Barrett, P. Fontaine, C. Tinelli, The SMT-LIB Standard: Version 2.6. Technical report, Department of Computer Science, The University of Iowa. Accessed 18 April 2024
2. A. Biere, Picosat essentials. J. Satisfiability, Boolean Model. Comput. **4**(2–4), 75–97 (2008)
3. A. Biere, A. Cimatti, E.M. Clarke, Y. Zhu, Symbolic model checking without BDDs, in *Tools and Algorithms for the Construction and Analysis of Systems* (1999), pp. 193–207
4. A. Biere, M. Fleury, F. Pollitt, CaDiCaL vivinst, IsaSAT, Gimsatul, Kissat, and TabularaSAT entering the SAT competition 2023, in *SAT COMPETITION 2023* (2023), p. 14

5. S. Chakraborty, K.S. Meel, M.Y. Vardi, A scalable and nearly uniform generator of SAT witnesses, in *Computer Aided Verification* (2013), pp. 608–623
6. E.M. Clarke, T.A. Henzinger, H. Veith, R. Bloem, et al. *Handbook of Model Checking*, vol. 10 (Springer, Berlin, 2018)
7. S.A. Cook, The complexity of theorem-proving procedures, in *Logic, Automata, and Computational Complexity: The Works of Stephen A. Cook* (2023), pp. 143–152
8. P. Davide Schiavone, F. Conti, D. Rossi, M. Gautschi, A. Pullini, E. Flamand, L. Benini, Slow and steady wins the race? A comparison of ultra-low-power RISC-V cores for internet-of-things applications, in *2017 27th International Symposium on Power and Timing Modeling, Optimization and Simulation* (2017), pp. 1–8
9. M. Davis, H. Putnam, A computing procedure for quantification theory. J. ACM **7**(3), 201–215 (1960)
10. M. Davis, G. Logemann, D. Loveland, A machine program for theorem-proving. Commun. ACM **5**(7), 394–397 (1962)
11. L. De Moura, N. Bjørner, Z3: an efficient SMT solver, in *Tools and Algorithms for the Construction and Analysis of Systems* (2008), pp. 337–340
12. B. Dutertre, Yices 2.2, in *Computer Aided Verification*, ed. by A. Biere, R. Bloem, vol. 8559. Lecture Notes in Computer Science (Springer, Berlin, 2014), pp. 737–744
13. N. Eén, N. Sörensson, An extensible sat-solver, in *International Conference on Theory and Applications of Satisfiability Testing* (Springer, Berlin, 2003), pp. 502–518
14. S. Horne, A. Bacherov, contributors. Openrisc isa. https://raw.githubusercontent.com/openrisc/doc/master/openrisc-arch-1.4-rev0.pdf. Accessed 18 April 2024
15. M.G. Katevenis, R.W. Sherburne, D.A. Patterson, C.H. Séquin, The RISC II micro-architecture. Adv. VLSI Comput. Syst. **1**(2), 138–152 (1984)
16. D.D. Lee, S.I. Kong, M.D. Hill, G.S. Taylor, D.A. Hodges, R.H. Katz, D.A. Patterson, A VLSI chip set for a multiprocessor workstation. I. An RISC microprocessor with coprocessor interface and support for symbolic processing. IEEE J. Solid-State Circuits **24**(6), 1688–1698 (1989)
17. J. Marques-Silva, K.A. Sakallah, Grasp: a search algorithm for propositional satisfiability. IEEE Trans. Comput. **48**(5), 506–521 (1999)
18. M.W. Moskewicz, C.F. Madigan, Y. Zhao, L. Zhang, S. Malik, Chaff: engineering an efficient sat solver, in *Proceedings of the 38th Annual Design Automation Conference* (2001), pp. 530–535
19. M.D. Nguyen, M. Thalmaier, M. Wedler, J. Bormann, D. Stoffel, W. Kunz, Unbounded protocol compliance verification using interval property checking with invariants. IEEE Trans. Comput. Aided Des. Circuits Syst. **27**(11), 2068–2082 (2008)
20. A. Niemetz, M. Preiner, A. Biere, Boolector 2.0. J. Satisf. Boolean Model. Comput. **9**(1), 53–58 (2014)
21. D.A. Patterson, C.H. Sequin, RISC I: a reduced instruction set VLSI computer, in *Proceedings of the 8th Annual Symposium on Computer Architecture* (IEEE Computer Society Press, Washington, 1981), pp. 443–457
22. PicoRV32 - A Size-Optimized RISC-V CPU. https://github.com/YosysHQ/picorv32. Accessed 18 April 2024
23. SMT-COMP 2024. https://smt-comp.github.io/2024/. Accessed 14 April 2024
24. SPARC International Inc. The SPARC Architecture Manual Version 8 (1992)
25. The pulp platform. https://www.pulp-platform.org/. Accessed 20 April 2024
26. D. Ungar, R. Blau, P. Foley, D. Samples, D. Patterson, Architecture of SOAR: smalltalk on a RISC. SIGARCH Comput. Archit. News **12**(3), 188–197 (1984)
27. VexRiscv: A FPGA friendly 32 bit RISC-V CPU implementation. https://github.com/SpinalHDL/VexRiscv. Accessed 18 April 2024
28. A. Waterman, K. Asanović, *The RISC-V Instruction Set Manual; Volume I: Unprivileged ISA*. SiFive Inc. and CS Division, EECS Department, University of California, Berkeley (2019)

Chapter 3
Processor Verification by Equivalent Program Execution

This chapter introduces a processor verification methodology named EPEX which is based on a new form of equivalence checking. The core idea of this methodology is that two equivalent but distinct programs must produce the same result. By using a formal model of the processor specification, a test program can be transformed into an equivalent program. This "broadening" of the test program will activate different control and data paths in the hardware thus improving the chance for uncovering bugs. We evaluated EPEX on several experiments, including mutation experiments and the measurement of the control path variation produced by EPEX.

3.1 Introduction

With the advent of the RISC-V ISA [32, 33] and the principles of open source arriving in the semiconductor industry, teams started to design their own processor or accelerator as RISC-V is a free and open ISA. The RISC-V ISA standard is maintained by the nonprofit RISC-V international. Moreover, configurability and extensibility were important design factors of the RISC-V ISA. However, while this opens up many opportunities, two major issues occur: (1) There is not much field-proven verification experience for RISC-V processors, and (2) the flexibility of the RISC-V ISA w.r.t. ISA extensions results in significant additional verification effort. During processor design, effective verification of the RTL implementation at the ISA level is a mandatory task. Here, instructions are checked on the complete processor, and this also includes initialization code/boot code, peripheral access, etc. Hence, at the ISA level, test programs (binaries) are needed which are simulated on the design. While they are still very frequently created manually, also methods for automatic generation of test programs have been proposed to allow for thorough verification. Prominent examples include approaches using constraint solving techniques, guided random generation, or approaches such as quick error

L. Klemmer, D. Große, *Versatile Hardware Analysis Techniques*,
https://doi.org/10.1007/978-3-031-83093-8_3

detection. Overall, however, significant effort is required to produce thorough test programs.

In this chapter, we present **EPEX** a novel approach for processor verification. Our approach is based on a new form of equivalence checking: Instead of comparing the architectural states of two models which execute the same program P, we *iteratively* perform two integrated steps for verification: (1) We replace a single instruction of P by one or more instructions; more formally, we *locally* determine for the next to be executed instruction P_i of P an equivalent instruction sequence $\hat{P}_{ij}$ of length j. For this task, we have created a *formal ISA model*[1] for which we constrain the current architectural state and next architectural state when executing P_i and eventually make use of an SMT solver. (2) We check that executing $\hat{P}_{ij}$ and P_i will produce equal architectural states on the *same* design.

For processor verification, various approaches aiming at the generation of test programs have been proposed. A strong focus was put on the separation of the test generator from the architecture description. This includes, for instance, approaches using constraint solving techniques [1, 5]. Katz et al. [18] presents an optimized test generation framework which propagates constraints among multiple instructions effectively. The test program generator of [6] uses a constraint-based coverage model describing execution paths of individual instructions. Alternative approaches integrate coverage-guided test generation via Bayesian networks [10] and other machine learning techniques [4, 17] as well as fuzzing [21]. However, these approaches are either not designed for RTL verification, require significant effort until thorough tests are generated, or do not target the RISC-V ISA.

Recently, several RISC-V specific verification approaches emerged. Official handwritten test suites are provided in [24, 27]. They are typically used as a starting point when implementing a RISC-V processor. A model-based test generation approach is the RISC-V *Torture Test* [29] framework. It uses Scala and generates tests based on randomized instruction sequence templates. Herdt et al. [15] is another model-based approach that employs a constraint-based specification for test generation. For verification at the Instruction Set Simulator (ISS) level, approaches leveraging coverage-guided fuzzing have been proposed recently [13, 14]. Googles' *RISC-V DV* [25] is another test generation approach that leverages SystemVerilog in combination with the Universal Verification Methodology (UVM) to continuously generate RISC-V instruction streams based on constrained-random specifications. In [16], an efficient cross-level testing approach for processor verification which generates an endless instruction stream on the fly during simulation has been proposed.

Besides methods for test generation, formal approaches aiming at RISC-V have also been developed. Notable are *riscv-formal* [26] and the *OneSpin 360 DV* RISC-V verification app [23] from Siemens EDA, which are both based on advanced model checking techniques. However, for using *riscv-formal*, the user has to implement the RISC-V formal interface, and *OneSpin 360 DV* is only

[1] Our formal ISA model is available at https://github.com/ics-jku/epex-formal-rv32-model.

commercially available. Beyond this, initially proposed for post-silicon verification only, Quick Error Detection (QED) has been enhanced in a recently proposed symbolic form called Symbolic Quick Error Detection (SQED). SQED targets pre-silicon verification of processors [30] and has been extended and applied to RISC-V processors [9]. As such, it checks whether the outputs produced by executing a particular instruction sequence match if the sequence is executed twice, assuming the inputs to the two sequences also match. For this check, SQED leverages BMC [3, 22]. While SQED offers several advantages as a formal technique, there are also some limitations: Single instruction checks have to be performed for each instruction, i.e., each instruction has to be completely formally verified on the core. In contrast, EPEX can uncover bugs even in a basic data path operation, since a comparison to other unrelated logic is possible. Moreover, SQED involves splitting the register file in half and using one half for the original instructions and the second half for a duplicated sequence of instructions. However, this imposes practical restrictions; for example, tests cannot use the full register range. Overall, while formal approaches can provide correctness guarantees, they are significantly more difficult to use than simulation-based methods and, due to their complexity and potential scalability issues, should be complemented by simulation-based methods.

Also in the formal context, solutions to formalize the RISC-V ISA semantics have been introduced, for example, [11, 28]. They allow leveraging theorem proving to reason about the RISC-V ISA semantics and generate simulation back ends. Unfortunately, even if mentioned, no formal ISA model in propositional logic is generated or available which we would have integrated in EPEX. Therefore, we created our own model.

This general procedure of EPEX provides improved processor verification, which is further strengthened by exploiting two major degrees of freedom coming with the EPEX principle: (1) We can substitute the next instruction with exactly one other instruction, or we can insert longer linked instructions sequences which will increase the variety of instructions influencing the test result. (2) When searching for an instruction replacement, we can guide the SMT-solver such that P and $\hat{P}$ will activate very different control and data paths of the processor implementation and hence enabling to compare typically unrelated logic/operations against each other. EPEX is complementary to other techniques, as existing tests can be broadened automatically, and the verification check is extended across different data and control paths.

In total, EPEX broadens a given test program automatically. EPEX increases the chance of detecting bugs. For example, the result of a compute instruction can be compared against the result of a jump which has to be identical per construction on the architectural states but checks very different micro-architectural states.

We show that EPEX can easily be used in a simulation-based verification environment. In a RISC-V case study using different core configurations of the well-known VexRiscv core, we demonstrate the bug-finding capabilities of EPEX. The approach has been published in [19].

This chapter is structured as follows: Our EPEX approach is introduced in Sect. 3.2. The experimental evaluation is presented in Sect. 3.3. Finally, the chapter is concluded in Sect. 3.4.

3.2 EPEX

In this section, we first describe the basic idea of our approach in Sect. 3.2.1. Then, we describe the EPEX flow in Sect. 3.2.2. Afterward, we detail the core ingredients in the remaining sections.

3.2.1 *Basic Idea*

When verifying a processor implementation at the ISA level, test programs (binaries) are needed. As the ISA of a processor defines the interface between the hardware and the software, processing an instruction of a test program transforms the architectural state s, i.e., the state visible to the programmer, to a new architectural state s'. We leverage this programmer's view abstraction at the ISA level as follows: We take an instruction of a test program P starting in s leading to the successor architectural state s'. Now we use a formal ISA model and automated reasoning (SMT) to determine a different instruction, which, when executed on s, results in the same successor architectural state s' as the original instruction. This process is repeated for every instruction of P. By this, we have transformed P to $\hat{P}$ according to the ISA specification and know that the behavior of both programs is equivalent on the architectural states. Hence, instead of classical equivalence checking where two models are compared on identical input, we now consider one processor implementation (which is instantiated twice), simulate the equivalent programs P and $\hat{P}$, respectively, and compare the architectural states along the execution. Consequently, we denote our novel approach for processor verification as Equivalent Program EXecution (EPEX).

Taking the idea further, EPEX can be extended naturally to search for instruction sequences instead of only a single instruction used as replacement to form $\hat{P}$. In general, EPEX offers the following advantages: (1) Existing tests can be broadened automatically, (2) The search can be guided to activate different control and data paths when determining an instruction replacement, and (3) General purpose software can be converted into enhanced test programs.

We now provide two concrete examples to illustrate EPEX and in particular the broadening of test programs. First, we consider the replacement of the current instruction with a single instruction. Thereafter, we provide an example using two instructions for the equivalent program transformation.

```
0996: ...
# assume t1 = 2, t2 = 2046, t3 = 0
1000: add t3, t1, t2 # t3 = t1 + t2 = 2048
1004: ...
```

(a) Excerpt of test program *P*

```
0996: ...
# assume t1 = 2, t2 = 2046, t3 = 0
1000: slli t3, t1, 10 # shift left t1 by 10
1004: ...
```

(b) Excerpt of equivalent test program $\hat{P}1$

```
0996: ...
# assume t1 = 2, t2 = 2046, t3 = 0
1000: j 1044 # set pc to 2044
1004: ...

2044: jal t3, -1040 # jump back and link t3 = 2044 + 4 = 2048
```

(c) Excerpt of equivalent test program $\hat{P}2$

Fig. 3.1 Example for broadening of a test program. (**a**) Excerpt of test program *P*. (**b**) Excerpt of equivalent test program $\hat{P}1$. (**c**) Excerpt of equivalent test program $\hat{P}2$

Example 3.1 Figure 3.1a shows an excerpt of the test program *P* together with the memory addresses where the instructions of *P* are located. Let us consider the add instruction to be processed in a given architectural state. This architectural state has been reached by executing all instructions up to and including address 0996. We only list the relevant subset of this architectural state, i.e., the registers $t1 = 2$, $t2 = 2046$, and $t3 = 0$. After the execution of `add`, the target register $t3$ has been set to 2048. In Fig. 3.1b, an equivalent program, as found by EPEX employing formal methods, is shown where the `add` instruction has been replaced by the shift left logical immediate. More precisely, the execution of `slli` shifts $t1$ by 10 and stores the result of 2048 into $t3$. By this, both architectural states, after executing up to and including `add` of *P* and up to and including `slli` of $\hat{P}1$, respectively, are identical. This example demonstrates that the compute instruction `add` can be replaced by another compute instruction (here `slli`) in this particular scenario/architectural state. By this, EPEX broadens an existing test program *P* in the sense that typically unrelated logic, i.e., in particular different control and data path operations are executed, and their results are finally compared.

Example 3.2 We now look at a more complex example. Let us consider the same test program *P* of Fig. 3.1a again. However, this time we are searching for an instruction replacement consisting of two instructions. The result using EPEX is shown in Fig. 3.1c. The determined test program uses two jump instructions and a free memory location. In detail, from address 1000, a relative jump (`j`) to address

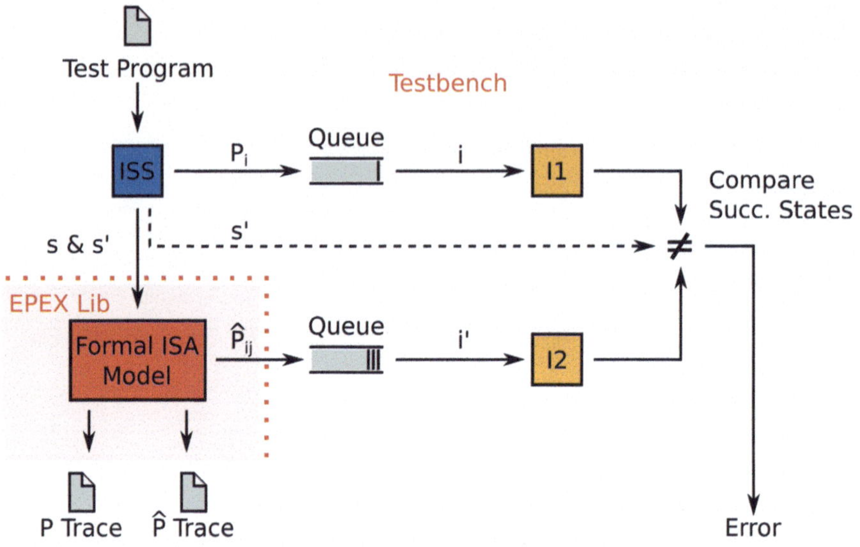

Fig. 3.2 Overview of the EPEX flow

2044 is performed. For this location, a jump-and-link instruction (`jal`) has been generated, which performs a relative jump to 1004 (the next instruction after the original add) and sets the target register $t3$ as the sum of the current program counter and 4, resulting in the needed value 2048. This new test program shows that an equivalent test program can be determined which now relates a compute instruction with two instructions from a completely different instruction class, i.e., control flow operations. These kinds of test programs will neither be created manually nor are random generators able to produce them reliably.

As both examples have demonstrated, EPEX opens up a novel approach for the verification of processors by broadening test programs. We introduce the EPEX flow in the next section.

3.2.2 EPEX Flow

The EPEX flow is depicted in Fig. 3.2. EPEX is split into two components: (1) The (processor) testbench controls the iterative execution of the original test program and drives the simulation of the two processor instances, denoted as $I1$ and $I2$ in Fig. 3.2, respectively. (2) The generation of equivalent instructions with the formal ISA model is performed via the EPEX library. This library is not tied to any particular processor implementation and can be easily connected to other cores.

EPEX starts with a user-supplied test program as input (top in Fig. 3.2). From the formal ISA model, we have also derived an ISS. Into this ISS, the test program is loaded by the testbench. Then, the ISS starts to execute the program step by step. For each step, the architectural states s and s' and the executed original instruction P_i are saved.

The formal ISA model is constrained such that the initial state of the model matches s and the final state matches s'. This constrained model is then given to the Z3 SMT-solver which finds assignments for the unconstrained instruction variables that lead from s to s'. The values of the instruction variables form an instruction sequence $\hat{P}_{ij}$ which is equivalent to the original instruction P_i. The instruction sequence $\hat{P}_{ij}$ consists, depending on additional constraints (see Sect. 3.2.4), of either a single instruction or multiple linked instructions. Both P_i and $\hat{P}_{ij}$ are pushed to instruction queues which are connected to the corresponding processor instances *I1* and *I2*, respectively. Moreover, P_i and $\hat{P}_{ij}$ are dumped into separate trace files by the EPEX library. This enables later testing via a cache mechanism without the cost of querying the formal model again.

The processor instances fetch their instructions not from memory but from the queues (gray in the center of Fig. 3.2). This allows controlling the instructions that are fetched and executed by *I1* and *I2*. After *I1* has finished the execution of P_i and *I2* has finished the execution of $\hat{P}_{ij}$, the architectural states of *I1*, *I2* and the *ISS* are compared. If the architectural states do not match, a bug either in *I1* or *I2* has been discovered, and the EPEX test is terminated and marked as failed. Recall that the ISS has been derived from the formal ISA model and is therefore correct by construction. However, *I1* and *I2* (instances of the processor at hand) execute different, but on the architectural states equivalent, programs as generated by EPEX. Therefore, in this case, the mismatch has to result from the activation of different control and/or data paths of the processor via the programs P_i and $\hat{P}_{ij}$, respectively.

In the following, we detail the core components, i.e., the formal ISA model and constraints for guidance of the search. Finally, the major implementation aspects are summarized.

3.2.3 *Formal ISA Model*

In this section, we introduce the formal ISA model which we have developed and which is used for the generation of equivalent programs.

Finding equivalent instruction sequences is a special case of the program synthesis problem [20] which is up to today a challenging task. Formal methods, such as SMT solving, are commonly used in program synthesis; see, for example, [12].

While multiple projects of formal RISC-V ISA specifications emerged, nothing is available which is suitable for automated reasoning. To minimize the effort for the creation of the formal ISA model, we decided to create it in C. This additionally allows to directly derive the ISS needed for EPEX. To integrate our formal model into EPEX, we use CBMC [7] to transform the C description into an SMT-lib v2.0 [2] instance. Hence, efficient reasoning becomes possible via the well-known SMT-solver Z3 [8].

To save run time in the EPEX generation phase, the SMT-lib instance produced by CBMC is post-processed to include placeholders for the architectural states s and s', respectively. During an EPEX run, these placeholders are replaced by the actual values of s and s'. Depending on the number of instructions that are to be generated, the model is unrolled. Each unwinding generates a new free 32-bit variable that encodes the instruction to be executed. Moreover, additional constraints to guide the search in case of generating an instruction sequence longer than one are necessary, which we discuss in the following section.

3.2.4 *Guidance Constraints*

If one instruction is replaced by an instruction sequence, the number of possible instruction sequences grows rapidly. Thus, some form of guidance is necessary, since many possible sequences even contain "uninteresting" instructions (e.g., different de facto NOP variations like `or x0, x0, x0`). To guide the formal search toward generating interesting instructions, we propose the following two constraints: The *effect constraint* specifies that the value of the destination register $s'[rd]$ in the successor architectural state s' is different from the value of the same register in the predecessor state s and by this reduces the number of de facto NOPs. Since not every instruction produces a result in a destination register (e.g., branches), the *effect constraint* is not always required. Overall, we get formally:

Definition 3.1 (Effect Constraint)

$$has\_result(current\_instr_p) \rightarrow s'[rd] \neq s[rd].$$

If instruction sequences are generated, the *chain constraint* guides the search to produce sequences with instructions that build on each other's results. This is reflected by relating the variables representing the registers along the generated instruction sequence P_{ij} with the index position p. Essentially, we formulate that consecutive instructions of the instructions sequence must use the result of the previous one, if one is produced:

Definition 3.2 (Chain Constraint)

$$has\_result(current\_instr_p) \rightarrow (rs1_{p+1} = rd_p \vee rs2_{p+1} = rd_p)$$

Besides these essential constraints, there are several options to drive the search in certain directions. We also conducted some experiments for this, but a thorough investigation is left for future work.

3.2.5 Implementation

As already shown in Fig. 3.2, EPEX consists of the EPEX library (containing the formal ISA model) and the processor testbench composed of the ISS, the processor instances, and the component for comparison of the architectural states. This separation makes the EPEX approach easily portable. The testbench needs to provide certain functionality to support EPEX. Explicit control over the fetched instructions is necessary. Moreover, it is essential that the exact time point a processor commits an instruction can be obtained. This is a common requirement for many verification approaches (e.g., [16, 26]).

3.3 Experimental Results

In this section, we present the experimental evaluation of our proposed EPEX approach. For the experiments, we use a modular RISC-V implementation for FPGAs, which is provided by the VexRiscv project [31]. The VexRiscv cores are built in SpinalHDL. SpinalHDL is a Domain Specific Language (DSL) in Scala which allows describing RTL by using object-oriented and functional programming. The VexRiscv cores are completely configurable from a bare-bones minimal CPU to a full-fledged Linux capable processor using a very flexible plugin concept. Finally, for a concrete core configuration, Verilog is generated.

In the first part of the experiments (Sect. 3.3.1), we reconsider the basic examples from Sect. 3.2. Next, we demonstrate the bug-finding capabilities of EPEX in a set of mutation-based experiments (Sect. 3.3.2). Finally, we show for multiple examples that executing P and $\hat{P}$ on the VexRiscv processor leads to a significant variation in the control paths.

3.3.1 Basic EPEX Examples

We come back to the test program excerpt described in Example 3.1. For this experiment, we selected a minimal VexRiscv configuration, but the goal was to speed up shifting of the core. During the design and integration of a barrel shifter in the RISC-V core, the shifting was split into three operations: In the case of a left shift operation, it first reverses the input. Then, it performs a right shift with variable number of steps. Finally, in the case of a left shift operation, it reverses the output again. Additionally, a configuration option was integrated to define the location of the final reverse operation to be either in the execute stage (reduces the logic depth) or in the memory stage of the pipeline. This has been modeled very nicely in SpinalHDL. However, due to a copy and paste error, the operation was not reversed correctly. Running our test program $\hat{P1}$ (Fig. 3.1b), which has been generated by EPEX, easily revealed this bug. Please note that it was immediately clear that the bug was not caused by the `add` instruction of the original program P (Fig. 3.1a) executed on instance $I1$: The comparison of the `add` result to the ISS gave the same output; the mismatch could be seen on the incorrectly computed register value $t3$ of the equivalent program $\hat{P1}$ (Fig. 3.1b) which executed the `slli` instruction on instance $I2$. Of course, dedicated shift tests would have also found this bug. However, the advantage of EPEX is that existing tests are *automatically* broadened and hence serve as an additional test.

Looking at the test program excerpt from Example 3.2 (`add` replaced by a `j` and a `rvasm`), and one of the first VexRiscv cores from the GitHub repository, we first observed no problem when using EPEX and comparing the execution of P and $\hat{P2}$, respectively. However, we instructed EPEX to generate more test programs. An alternative test program $\hat{P3}$ contained: `1000: jalr zero, 2045(zero)`.

For this program, EPEX reported an error. The reason is that the `jalr` instruction adds an immediate to its source register, then has to clear the LSB of the sum, and then jumps to the resulting address. In the VexRiscv implementation, however, the LSB was not set to zero and hence the program counters differed. The problem occurs when odd addresses are generated via EPEX. This bug is nowadays well-known in the RISC-V community and has been found by *riscv-formal* [26] on the VexRiscv core at hand. Nonetheless, this demonstrates that EPEX can be used to generate relevant tests via relating a compute instruction with two instructions from an entirely different instruction class, namely, control flow operations in this example.

3.3.2 Mutation Experiments

To evaluate the bug-finding capabilities of EPEX, we performed a set of mutation experiments which we describe in the following.

3.3.2.1 Experimental Setup

In this experiment, we used an area efficient VexRiscv configuration. We automatically created mutations of the Verilog code of the core. Mutants were created by toggling bit constants and modifying comparison operators (e.g., replacing greater than by greater than or equal). Overall, this resulted in 275 mutants.

3.3.2.2 EPEX Generation and Results

We now consider three tests of the VexRiscv test suite. For each test, EPEX needs on average about 5 minutes to generate an equivalent program. The obtained results for our approach are summarized in Table 3.1. The first column reports the name of

Table 3.1 Results for VexRiscv tests

Test case	#Instrs	Orig	EPEX
ADDI	263	81	87
SRAI	234	84	88
JALR	218	84	93

the test. The second column gives the number of instructions. In the third column, the number of mutants which have been discovered by the original test is reported. The last column reports the number of mutants identified with EPEX. As can be seen, EPEX can find nearly 8% more mutants than the original test on average. To achieve this result, we configured EPEX to replace each instruction with up to three alternative instructions for the generation of the equivalent test program.

3.3.2.3 EPEX Test Broadening Result

We found a very interesting mutant, which was not detected even with extended cross-level testing (includes full compliance test suite, Dhrystone, and CoreMark). In contrast to this, EPEX was able to discover the mutant by using the three EPEX tests shown Table 3.1. This demonstrates that broadening given tests using EPEX already increases the test quality even when using a few basic tests only.

3.3.2.4 EPEX Test Diversity

To illustrate the diversity resulting from the EPEX test programs, we look at the RISC-V instruction format. It is divided into six classes:[2] *R*egister/register, *I*mmediate, *U*pper immediate, *S*tore, *B*ranch, and *J*ump. The instruction format has been carefully designed to share and save logic. Moreover, it is important to note that all instructions of a class share most of the data path and have an almost identical control. This is because the instructions within a class use the same number of registers, immediates, etc. but, of course, differ in the specific operation. Based on this classification, we now illustrate the effect of EPEX when generating an equivalent test program. For each class, the percentage of replacements into other classes for the ADDI test is shown in Fig. 3.3.

For example, consider the class U (on the x-axis the fourth instruction format class): As can be seen, an instruction from this class is replaced in 10% of the cases with an instruction from J, in 15% of the cases with an instruction from B, in 0% of the cases with an instruction from S, in 17% of the cases with another instruction within this class, in 58% with an instruction from I, and in 0% of the cases with an instruction from R. Please note that instructions from the class S are intentionally

[2] See page 16, Figure 2.3 of "The RISC-V Instruction Set Manual Volume I: Unprivileged ISA V20191213" [32].

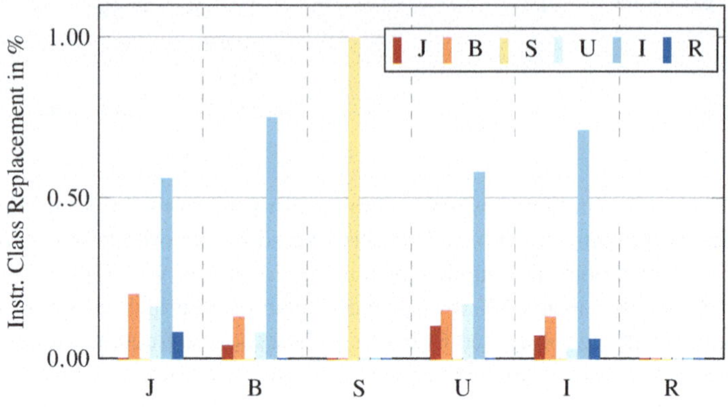

Fig. 3.3 Exemplary distribution of instruction class replacements for ADDI test (cf., Table 3.1)

Table 3.2 EPEX control path variation on VexRiscv

Test	EPEX	#Instrs	Gen ratio	Avg. Δ	Max. Δ
Compliance	1	8,834	100%	7.2%	36.0%
Compliance	1–4	12,923	100%	15.1%	45.3%
Dhrystone	1	38,798	10%	10.8%	33.0%
Dhrystone	1–4	41,931	10%	13.4%	38.0%

never replaced as integrating a memory model into the formal model would make it much harder to solve for the SMT solver. Furthermore, the ADDI test does not contain any instructions for class R. Overall, this plot demonstrates that a good diversity in the sense of replacing an instruction with an instruction from another class can be achieved. Therefore, independent of a RISC-V core implementation, diverse testing becomes possible using EPEX.

3.3.3 *Control Path Variation*

Finally, we consider the control path variation produced by EPEX. We used the same experimental setup as in the previous section. As tests, we utilized the compliance test suite and the Dhrystone benchmark. Table 3.2 lists the obtained results. The first column shows the name of the test. Column *EPEX* reports the number of instructions we configured to be replaced during equivalent program generation via the formal ISA model for an instruction. The column *#Instrs* shows the number of executed instructions. Then, column *Gen Ratio* gives the ratio of instructions of P that were selected for EPEX replacement. The next column *Avg.* Δ reports the average control path variation (stated in percent) triggered by $\hat{P}$ in comparison to P. For this, we carefully identified the select signals of the multiplexers which control the data path in the VexRiscv implementation along the pipeline stages. In the last column *Max.* Δ, the maximum delta found is provided, respectively. As can be seen, EPEX causes a strong variation of the control paths of up to 45%. The obtained percentage values clearly demonstrate that large parts of the control paths differ during the execution of P on instance *I1* and $\hat{P}$ on instance *I2*, respectively. Overall, the strong control path variation shows that EPEX compares typically unrelated logic/operations against each other which enables a wide testing.

3.4 Summary

In this chapter, we presented EPEX, a novel approach for processor verification that automatically broadens existing test programs. For this task, a formal ISA model and SMT reasoning are utilized to generate an equivalent test program. EPEX is based on a new form of equivalence checking: Executing both programs, i.e., original and equivalent, iteratively on two instances of the same processor design has to always result in equal architectural states. We demonstrated the bug-finding capabilities of EPEX on the VexRiscv processor. Broadening a few tests already resulted in the detection of more bugs.

EPEX is a promising new approach to processor verification. In particular, the feature of relating the execution of instructions which activate different control and data paths has great potential. For future work, we will investigate the replacement of longer instruction sequences in more detail. Furthermore, we will devise new guidance constraints in combination with novel coverage metrics to measure the difference in logic activity when running both programs on the two processor instances. Finally, we want to shift more of the EPEX flow, in particular the equivalence, check offline, thus enabling the execution of more instructions and the application of EPEX on performance-constrained devices.

References

1. A. Adir, E. Almog, L. Fournier, E. Marcus, M. Rimon, M. Vinov, A. Ziv, Genesys-pro: innovations in test program generation for functional processor verification. IEEE Des. Test Comput. **21**(2) 84–93 (2004)
2. C. Barrett, P. Fontaine, C. Tinelli, The SMT-LIB Standard: Version 2.6. Technical report, Department of Computer Science, The University of Iowa. Accessed 18 April 2024
3. A. Biere, A. Cimatti, E.M. Clarke, Y. Zhu, Symbolic model checking without BDDs, in *Tools and Algorithms for the Construction and Analysis of Systems* (1999), pp. 193–207
4. N. Bruns, D. Große, R. Drechsler, Early verification of ISA extension specifications using deep reinforcement learning, in *ACM Great Lakes Symposium on VLSI* (2020), pp. 297–302
5. B. Campbell, I. Stark, Randomised testing of a microprocessor model using SMT-solver state generation, in *Formal Methods for Industrial Critical Systems* (2014), pp. 185–199
6. M. Chupilko, A. Kamkin, A. Kotsynyak, A. Tatarnikov, Microtesk: specification-based tool for constructing test program generators, in *Hardware and Software: Verification and Testing* (Springer International Publishing, Berlin, 2017), pp. 217–220
7. E. Clarke, D. Kroening, F. Lerda, A tool for checking ANSI-C programs, in *Tools and Algorithms for the Construction and Analysis of Systems* (2004), pp. 168–176

8. L. De Moura, N. Bjørner, Z3: an efficient SMT solver, in *Tools and Algorithms for the Construction and Analysis of Systems* (2008), pp. 337–340
9. K. Devarajegowda, M.R. Fadiheh, E. Singh, C.W. Barrett, S. Mitra, W. Ecker, D. Stoffel, W. Kunz, Gap-free processor verification by s^2qed and property generation, in *Design, Automation and Test in Europe* (2020), pp. 526–531
10. S. Fine, A. Ziv, Coverage directed test generation for functional verification using bayesian networks, in *Design Automation Conference* (2003), pp. 286–291
11. Formal specification of RISC-V ISA in Kami. https://github.com/sifive/RiscvSpecFormal. Accessed 18 April 2024
12. S. Gulwani, S. Jha, A. Tiwari, R. Venkatesan, Synthesis of loop-free programs. SIGPLAN Not. **46**(6), 62–73 (2011)
13. V. Herdt, D. Große, H.M. Le, R. Drechsler, Verifying instruction set simulators using coverage-guided fuzzing, in *Design, Automation and Test in Europe* (2019), pp. 360–365
14. V. Herdt, D. Große, R. Drechsler, Closing the RISC-V compliance gap: looking from the negative testing side, in *Design Automation Conference* (2020), pp. 1–6
15. V. Herdt, D. Große, R. Drechsler, Towards specification and testing of RISC-V ISA compliance, in *Design, Automation and Test in Europe* (2020), pp. 995–998
16. V. Herdt, D. Große, E. Jentzsch, R. Drechsler, Efficient cross-level testing for processor verification: a RISC-V case-study, in *Forum on Specification and Design Languages* (2020), pp. 1–7
17. C. Ioannides, G. Barrett, K. Eder, Feedback-based coverage directed test generation: an industrial evaluation, in *Proceedings of the 6th International Conference on Hardware and Software: Verification and Testing* (2011), pp. 112–128
18. Y. Katz, M. Rimon, A. Ziv, Generating instruction streams using abstract CSP, in *Design, Automation and Test in Europe* (2012), pp. 15–20
19. L. Klemmer, D. Große, EPEX: processor verification by equivalent program execution, in *ACM Great Lakes Symposium on VLSI* (2021), pp. 33–38
20. Z. Manna, R.J. Waldinger, Toward automatic program synthesis. Commun. ACM **14**(3), 151–165 (1971)
21. L. Martignoni, R. Paleari, G.F. Roglia, D. Bruschi, Testing CPU emulators, in *Proceedings of the Eighteenth International Symposium on Software Testing and Analysis* (2009), pp. 261–272
22. M.D. Nguyen, M. Thalmaier, M. Wedler, J. Bormann, D. Stoffel, W. Kunz, Unbounded protocol compliance verification using interval property checking with invariants. IEEE Trans. Comput. Aided Des. Circuits Syst. **27**(11), 2068–2082 (2008)
23. OneSpin 360 DV RISC-V Verification App. https://www.onespin.com/solutions/risc-v. Accessed 18 April 2024
24. RISC-V compliance task group. https://github.com/riscv/riscv-compliance. Accessed 18 April 2024
25. RISC-V DV. https://github.com/chipsalliance/riscv-dv.
26. RISC-V formal verification framework. https://github.com/SymbioticEDA/riscv-formal. Accessed 18 April 2024
27. RISC-V ISA tests. https://github.com/riscv/riscv-tests. Accessed 18 April 2024
28. RISCV sail model. https://github.com/rems-project/sail-riscv. Accessed 18 April 2024
29. RISC-V torture test generator. https://github.com/ucb-bar/riscv-torture. Accessed 18 April 2024
30. E. Singh, D. Lin, C.W. Barrett, S. Mitra, Logic bug detection and localization using symbolic quick error detection. IEEE Trans. Comput. Aided Des. Circuits Syst. (2018)
31. VexRiscv: A FPGA friendly 32 bit RISC-V CPU implementation. https://github.com/SpinalHDL/VexRiscv. Accessed 18 April 2024
32. A. Waterman, K. Asanović, *The RISC-V Instruction Set Manual; Volume I: Unprivileged ISA*. SiFive Inc. and CS Division, EECS Department, University of California, Berkeley (2019)
33. A. Waterman, K. Asanović, *The RISC-V Instruction Set Manual; Volume II: Privileged Architecture*. SiFive Inc. and CS Division, EECS Department, University of California, Berkeley (2019)

Chapter 4
A Formally Verified Microcoded RISC-V Platform

This chapter presents a fully microcoded RISC-V platform based on the One Instruction Set Computer (OISC) principle. It combines the efficiency of OISC architectures with the established and extensive RISC-V ecosystem. The platform is introduced in two steps. First, a Virtual Prototype (VP) that facilitates the development of the OISC microcode is presented. Then, the microcode is verified using a formal verification framework based on a model of the OISC architecture.

4.1 Introduction

Besides its openness, major advantages of RISC-V include its modularity, extensibility, and the community-driven ecosystem. Due to the implementation-independent specification of the RISC-V ISA, RISC-V processors targeting many use cases are developed, i.e., ranging from high performance to microcontrollers which are crucial for the IoT domain. In the latter, strict constraints on energy consumption and area are inevitable. To reduce the area footprint to the minimum, typically performance is traded for a much more compact design.

A very promising approach in this direction, and seen from the RISC paradigm an extreme case, is an instruction set with a single instruction only. The idea to reduce the number of instructions of a RISC to the minimum, i.e., to a single instruction, led to the ultimate RISC computer or One Instruction Set Computer (OISC) [12, 15]. The careful selection of the instruction makes an OISC Turing-complete; i.e., it can solve any computing problem. Depending on the selected instruction, three OISC types can be distinguished:

1. *Arithmetic-based architectures:* This type uses an arithmetic operation and a conditional jump. A well-known example of this class is the *SUBLEQ* instruction, which subtracts and branches unless the result is positive [18].

L. Klemmer, D. Große, *Versatile Hardware Analysis Techniques*,
https://doi.org/10.1007/978-3-031-83093-8_4

2. *Bit-manipulating architectures:* These machines perform bit operations, like bit-flipping or copying, and pass the control flow either conditionally or unconditionally. Due to their extreme minimalism, they are the least practical relevant OISC type [17].
3. *Transport-triggered architectures:* While the only available instruction is *MOVE*, arithmetic, control flow, or other operations are available by writing to memory mapped registers; see, e.g., [6, 11].

The strengths of OISC machines lie in their extremely small area footprint and their high flexibility w.r.t. very specific use cases. Thus, OISC machines have been considered in a wide variety of applications, e.g., fault detection [3, 21], cryptography [8], stream processing [23], carbon nanotube computer [22], and (advanced) microcontrollers [6, 16].

However, programming directly in OISC languages is extremely complicated due to the limited features resulting in exceptionally long programs that are hard to debug and maintain. Further, most of these architectures require customized software tooling, in particular from the compiler side. This is because typically each OISC machine uses a custom ISA, thus hindering the emergence of a common OISC ecosystem. Yet, the availability of a mature and widely used hardware and software ecosystem is one of the deciding factors for the success of most technologies.

In this chapter, we present a fully microcoded RISC-V platform based on the OISC principle that combines the efficiency of OISC architectures with the established and extensive RISC-V ecosystem. We start this chapter with an introduction to the *SUBLEQ* OISC architecture in Sect. 4.2. Since implementing a fully specification-compliant implementation of RISC-V based on an OISC instruction is nontrivial, we took two steps in the development process. First, we created a VP to serve as a development platform for the microcode and to help guide architectural decisions during design space exploration. We present the VP in Sect. 4.3 and evaluate it in Sect. 4.4. Then, we constructed a formal model of the RISC-V specification and of the OISC architecture to enable formal verification of our microcode in Sect. 4.5. In Sect. 4.6, we present the results of verifying our microcode with the proposed formal model. Finally, we conclude this chapter in Sect. 4.7. This chapter is based on two conference papers [13, 14].

4.2 The SUBLEQ Instruction

Since it is quite intuitive, SUBLEQ is one of the best studied OISC instructions. The SUBLEQ instruction is a three-operand instruction, performing a combined subtraction and branching operation. The semantic of *SUBLEQ A B C* is as follows:

$$
\begin{aligned}
r &\leftarrow reg[B] - reg[A] \\
reg[B] &\leftarrow r \\
pc &\leftarrow \begin{cases} pc + C, & \text{if } r \leq 0 \\ pc + 1, & \text{otherwise} \end{cases}
\end{aligned}
$$

First, the result r is calculated by subtracting the register value in A from the register value in B. The result is written back in register B, and a jump is performed depending on the value of r: If r is smaller or equal to 0, C is added to the SUBLEQ program counter, else the SUBLEQ program counter is incremented by 1.

To ease the development of SUBLEQ procedures, we implemented a SUBLEQ file format and an assembler that translates assembled SUBLEQ programs to multiple output formats, e.g., Racket lists or Verilog memory dumps. This assembler resolves jump labels to concrete offsets and also creates definitions for the entry points of all SUBLEQ procedures. For the *JAL* RISC-V instruction, we show a SUBLEQ procedure in this format in the following example.

Example 4.1 Even though the *SUBLEQ* instruction only provides a way to subtract one number from another, complex RISC-V instructions, such as ***JAL***, can be implemented. Listing 4.1 shows the SUBLEQ procedure for the *JAL* instruction. The *jal rd, imm* instruction performs a *Jump And Link* operation, i.e., adding the immediate value *imm* to the current program counter and storing the current program counter + 4 in the destination register *rd*. At the start of the execution, the *JAL* procedure expects the immediate value *imm* to be present in SUBLEQ register `IMM`.[1] First, the procedure resets the temporary registers TMP0 and TMP1 as well as the result register `RSLT` to 0 by subtracting them from themselves (Lines 2–4). Since this is a common operation in SUBLEQ code, it can be abbreviated by writing just one register in a line using our assembler. It is important to reset the registers, since they still can contain garbage data from previous procedures which would corrupt the results of the current procedure.

```
1 @JAL
2    TMP0           ; reset TMP0
3    TMP1           ; reset TMP1
4    RSLT           ; reset RSLT
5    RVPC TMP0      ; -pc => TMP1
6    TMP0 RSLT      ; pc => SRC2
7    NEXT RSLT      ; add link address +4
8    IMM  TMP1      ; negate IMM
9    TMP1 RVPC END  ; PC - -TMP0 link1
```

Listing 4.1 SUBLEQ implementation of the JAL instruction

The procedure first handles computing the link address (RISC-V pc + 4). For this, the current RISC-V program counter in `RVPC` has to be copied to the `RSLT` register. To achieve this, the `RVPC` is negated and written into `TMP0` by subtracting it

[1] This will be managed later by the RISC-V interface.

from register `TMP0` (which contains a 0 at this point, Line 5). Next, the pc is negated again and copied into the `RSLT` register (Line 6). Register `NEXT` contains the constant −4 which is subtracted from the `RSLT` register to calculate the link address (Line 7).

What remains now is to add the immediate value containing the jump offset to the RISC-V pc. This is done similarly by first negating the value and subtracting this negated value from the `RVPC` register (Line 8). Since this is the last operation of this procedure, the jump target is set to `END` which stops the SUBLEQ execution (Line 9).

4.3 ORISCV-ISS: A RISC-V Virtual Prototype Using SUBLEQ Microcode

In modern industry-proven design flows, VPs are created to enable the parallel development of hardware and software in combination with design space exploration [4, 5, 7, 10]. A VP is essentially an executable abstract model of the entire hardware platform, and from the software perspective, a VP mimics the real hardware. For enabling the execution of programs by the VP, an Instruction Set Simulator (ISS) has to be created for the processor. We follow this approach and take advantage of the ISS principle and by this significantly enhance the design flow of OISCs.

In this section, we present an exploration platform for microcoded RISC-V cores leveraging the OISC principle. Following the industry-proven virtual prototyping approach, we have realized our exploration platform by implementing an extensible and configurable ISS. The developed ORISCV-ISS combines the advanced ecosystem of RISC-V with the ultimate minimalism of OISCs. ORISCV-ISS serves as a development platform for both, hardware architecture and microcode procedures, and provides the basis for early design space exploration.

Using ORISCV-ISS, we developed *SUBLEQ* microcode that is fully RISC-V compliant and ready to be run on real hardware. We evaluate how multiple hardware configurations and OISC extensions affect the performance, providing key information to balance between area savings and system performance.

ORISCV-ISS serves as a development platform for both, hardware architecture and microcode procedures, and provides the basis for early design space exploration. This is demonstrated by utilizing the *SUBLEQ* OISC architecture to serve as the microcode basis: From the software side, ORISCV-ISS accepts RISC-V code; i.e., it behaves like an RV32I compliant core.[2] However, ORISCV-ISS decodes each RISC-V instruction and selects the defined *SUBLEQ* microcode procedure to be executed on the OISC execution unit. To enable design space exploration, we have integrated a timing model into ORISCV-ISS. Hence, we can evaluate the effect of various memory configurations (which define the location of the different

[2] ORISCV-ISS passes all official RISC-V compliance tests.

registers, such as RISC-V registers, OISC registers, and functional registers) on the performance. At the same time, ORISCV-ISS enables the parallel development and exploration of microcode procedures w.r.t. alternatives in the hardware support of the chosen OISC instruction. Altogether, ORISCV-ISS provides key information to balance between area savings and system performance in an early design stage while leveraging the benefits of the RISC-V ecosystem.

4.3.1 ORISCV-ISS *Architecture*

In this section, we introduce our proposed ORISCV-ISS. First, we give an overview about the ORISCV-ISS and discuss the architecture. Then, in the following sections, we describe the main components in detail.

4.3.1.1 ORISCV-ISS Overview and Architecture

ORISCV-ISS has been implemented in approximately 1,000 lines of plain C++ code. The current microcode implementation(s) support RV32I, although the ISS itself features all requirements to support additional RISC-V extensions. An overview of the architecture of ORISCV-ISS is shown in Fig. 4.1. The ISS core has been structured into four main components: (1) the main memory, (2) the RISC-V interface, (3) the OISC execution unit, and (4) the timing model. In Fig. 4.1 and the rest of this chapter, all components (or data flows) related to RISC-V are colored in blue, and all components (or data flows) related to the OISC microcode are colored in red. In the following sections, we describe all components in more detail.

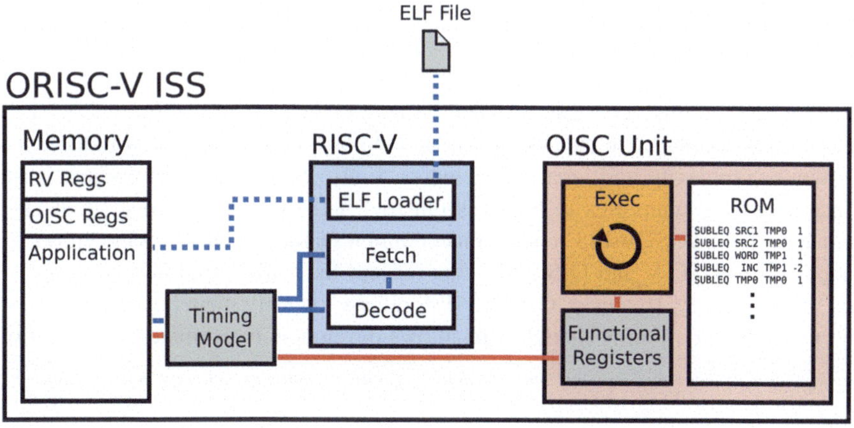

Fig. 4.1 Architecture of the ORISCV-ISS

4.3.1.2 RISC-V Interface

To the user, ORISCV-ISS is a RISC-V compliant ISS, completely hiding the microcode layer. Internally, the *RISC-V interface* (center of Fig. 4.1 with blue background) handles loading RISC-V binaries, and fetching plus decoding RISC-V instructions. Before the ISS starts executing a RISC-V binary, it is loaded into the main memory using the ELF loader (see blue dotted line). Then, the ELF loader sets the initial RISC-V program counter and starts decoding the first RISC-V instruction. After the RISC-V interface has decoded an RISC-V instruction, it sets up the OISC execution unit by first putting register values and immediate values of the RISC-V instruction at hand into the OISC registers. It then looks up the address of the microcode implementation for the current RISC-V instruction and hands over the execution to the OISC unit. After the OISC unit finishes its computation, the result, if one is produced, is loaded from the OISC registers and stored back at the RISC-V destination register. Then, the next RISC-V instruction is processed.

4.3.1.3 OISC Execution Unit

The OISC execution unit (right side in Fig. 4.1) implements the execution loop of the OISC microcode.[3] During execution, microcode instructions are fetched from a *very small* microcode ROM (less than 1 KB), executed, and then their results are written back to the OISC registers. Moreover, microcode instructions can perform additional operations (e.g., bit-operations) by writing to memory-mapped *functional registers*. Using these registers, developers can integrate new operations and evaluate their performance impact. By this, ORISCV-ISS allows a hybrid OISC design space exploration of a mixture of arithmetic-based and transport-triggered OISC architectures.

4.3.1.4 Memory

Often, OISC architectures are designed without dedicated hardware registers; i.e., all registers are placed in the system's main memory. While this greatly reduces the system's complexity and area requirements, it also has a high-performance cost, since accessing memories is highly sequential and slow compared to dedicated registers. The ORISCV-ISS reflects this common feature of OISC architectures by mapping all RISC-V and OISC registers (except the OISC program counter) into the main memory. However, ORISCV-ISS allows to easily model access times on the registers via a timing model. This allows developers to estimate the impact of

[3] As already mentioned, we use *SUBLEQ* microcode instructions, but ORISCV-ISS can easily be extended to other OISC instructions.

the system architecture on the runtime without modification of the microcode. We will use this feature intensively in the experiments in Sect. 4.4.

4.3.1.5 Timing Model

To enable design space exploration of different architectures (e.g., dedicated registers vs. registers in memory), every access to the memory (except from the ELF loader since it runs before simulation time) is routed through the timing model. In Fig. 4.1, the timing model is located between the memory and the RISC-V interface. In the timing model, developers can specify the time required to access different registers or memory regions. The timing model adds a delay to the simulation for every read or write operation on the main memory and can be customized by the user. This way, even though all registers (RISC-V and OISC) are contained in the main memory, the timing model enables the evaluation of execution metrics when running RISC-V software. For example, the added delay can be significantly reduced for reads or writes to the OISC registers if they are implemented in dedicated registers. Naturally, these dedicated registers incur significant area costs when implemented in hardware. To summarize, the timing model enables early design space exploration, thus helping developers to decide on the hardware architecture.

In the next section, the microcode layer of ORISCV-ISS is introduced, and the RISC-V compliance is shown.

4.3.2 Microcode Layer and RISC-V Compliance

In this section, we introduce the microcode layer of ORISCV-ISS. This includes in particular the registers visible to the microcode, as well as the microcode instruction format. Next, we show a concrete microcode example for an RISC-V instruction. Finally, we show the RISC-V compliance of ORISCV-ISS.

4.3.2.1 Registers and Microcode Instruction Format

Since compactness is a major goal during the design of microcode, we adapt the instruction format of *SUBLEQ A B C*: In comparison to other OISC architectures, the operands *A* and *B* do not have to address the full main memory, but only a limited number of registers. Since in other works applications are directly compiled to OISC instructions, they must be able to address the complete memory. In contrast, for ORISCV-ISS, only a few registers are sufficient to implement the RISC-V instructions. Limiting the *A* and *B* operands of the *SUBLEQ* instruction

Table 4.1 16 OISC registers visible to the microcode. Some registers have two names, because they serves two purposes

Operands	Temporary	Constants	Functional
SRC1	TMP0	ONE	FUNC0/TMP6
SRC2/RSLT	TMP1	TWO	FUNC1/TMP7
IMM	TMP2	WORD	
	TMP3	INC	
	TMP4	NEXT	
	TMP5		

to a few registers only enables us to reduce the instruction length, thus saving a precious amount of microcode memory. Table 4.1 lists the registers available to the microcode.

The registers can be grouped into four categories: (1) operands for passing the RISC-V operands; (2) temporary registers; (3) constants that store frequently needed values, such as -1 (in register `INC`) to increment by one, or -4 (in register `NEXT`) to increment the RISC-V program counter; and (4) functional registers for additional hardware support of specific operations. Usually, the *SUBLEQ A B C* instruction sets the OISC program counter to the value of operand *C*, if the resulting value r of the *SUBLEQ* computation is smaller or equal to 0. This allows to perform long-range jumps, which are crucial for programming applications directly in the OISC language. However, in the proposed microcode, only local jumps of a few instructions are needed, enabling us to change the global jump target to a pc-relative jump. This again allows reducing the instruction length, saving a lot of space. We provide a concrete example demonstrating the considerations.

Example 4.2 For this example, we assume a main memory of 16 kB. Fully addressing this memory requires 14 bits. Therefore, one instruction has a length of $3 * 14 = 42$ bits.[4] With the above introduced enhancements and the microcode registers of Table 4.1, we only need 4 bits to address every register. Additionally, we assume that no microcode jump is more than a generous 128 instructions long, requiring 8 additional bits for the signed jump offset operand. In total, this results in a $4 + 4 + 8 = 16$ bit long microcode instruction, thus reducing the length by 60%.

4.3.2.2 *SUBLEQ* Microcode Implementing RV32I

For the implementation of the RV32I instructions as *SUBLEQ* microcode procedures, we required no additional hardware support.

In the following, we show exemplary microcode procedures for the RISC-V immediate addition *addi* and shift left logical *sll*, respectively.

[4] We also assume that the program resides in the main memory as in most OISC implementations.

Example 4.3 Consider Listing 4.2. Before the microcode procedure realizing the RISC-V instruction *addi t0, t1, 22* is started,[5] the RISC-V interface copies the required operands from RISC-V register `t1` and the immediate value 22 into the OISC registers `SRC2` and `IMM`, respectively. Then, on Line 1, the *22* in `IMM` is subtracted from OISC register `TMP0`, and the result is stored back in `TMP0`. Since `TMP0` is guaranteed to contain a zero at the start of the microcode procedure, OISC register `TMP0` now holds the value *-22*. Next, `TMP0` is subtracted from `SRC2` which finally performs the addition via a doubled negation. At the end of the procedure, the temporary register `TMP0` is cleared (Line 3), and the RISC-V program counter is incremented (Line 4). Finally, the RISC-V interface reads the result from `SRC2` and copies it into the destination register `t0`.

```
addi t0, t1, 22
# RISC-V Interface: reg[SRC2] = reg[t1]
# RISC-V Interface: reg[IMM] = 22

1  IMM  TMP0    1                  # TMP0 = -22
2  TMP0 SRC2    1                  # SRC2 = SRC2 - (-22)
3  TMP0 TMP0    1                  # TMP0 = 0
4  NEXT   PC  END                  # RV-PC = RV-PC + 4

# RISC-V Interface: reg[t0] = reg[SRC2]
```

Listing 4.2 Microcode procedure for RISC-V addi

Example 4.4 Typical *SUBLEQ* shift operations make use of the fact that a single left shift of x is equal to $x + x$ and simply repeating this operation depending on the shift amount. Listing 4.3 contains the microcode procedure that implements the `sll` RISC-V instruction. Before the execution of a `sll` instruction, the RISC-V interface places the shift amount in register `SRC1` and the value to be shifted in `SRC2`. First, on Line 1, the shift amount is negated and stored into register `TMP0`. Then, the shift loop body is entered in which a left shift of one bit is performed by adding the current shift value in `SRC2` to itself on Line 2–3. At the end of the shift loop body (Line 4), the temporary register `TMP1`, which stored the negated shift value, is cleared. On Line 5, we increment the negated shift amount in register `TMP0` and jump back to the beginning of the shift loop body if the result is smaller or equal to 0. Note that in *SUBLEQ* microcode loops, counters are typically counting from negative to positive, since this can easily be implemented using the *SUBLEQ* instruction. If the loop is left, the used temporary registers are cleared on Line 6–7 and the RISC-V program counter is incremented on Line 8. Finally, the RISC-V interface reads the result of the shift operation from `SRC2` and writes it into the destination register `t1`.

```
sll t0, t1, t2
# RISC-V Interface: reg[SRC1] = reg[t2]
# RISC-V Interface: reg[SRC2] = reg[t1]
```

[5] We use a concrete immediate value of 22 in the example for ease of understanding.

```
SRC1 TMP0 2                         # TMP0 = -SRC1
SRC2 TMP1 1                         # TMP1 = -SRC2
TMP1 SRC2 1                         # SRC2 = SRC2 - (-TMP1)
TMP1 TMP1 1                         # TMP1 = 0
INC  TMP0 -3                        # TMP0 = TMP0 + 1
TMP0 TMP0 1                         # TMP0 = 0
TMP1 TMP1 1                         # TMP1 = 0
NEXT PC   END                       # RV-PC = RV-PC + 4

# RISC-V Interface: reg[t0] = reg[SRC2]
```

Listing 4.3 Microcode procedure for RISC-V sll

In the next section, we show the RISC-V compliance of the developed microcode.

4.3.2.3 RISC-V Compliance of Microcode

The *RISC-V Architectural Tests* from the RISC-V foundation provide a standardized testing infrastructure for checking that a RISC-V implementation (ISS or hardware) meets the RISC-V ISA standard [1]. The architectural test suite is split up in multiple tests for each RISC-V configuration. To check that ORISCV-ISS correctly implements the RV32I configuration, we created a new target architecture for our ISS and compiled the RV32I test suite. Overall, the RV32I base instruction set contains 37 unprivileged instructions. Our microcode implementation passes the complete RV32I architectural tests for each of the 37 instructions. In sum, the architectural test suite consists of 12,603 individual test cases.

4.3.2.4 SUBLEQ Correctness and Fixed Word Lengths

During the development of ORISCV-ISS as well as the OISC microcode procedures for each RISC-V instruction, we also looked into the available OISC literature. While the idea of mapping a certain operation to *SUBLEQ* microcode is pretty simple, corner cases in particular w.r.t. to overflows can be problematic.

```
1 SRC1 TMP0 2                       # 0 - (-4) = -4
2 TMP1 TMP1 END                     # correct
3 TMP1 TMP1 -5                      # wrong
```

Listing 4.4 Overflow error leading to wrong branch taken

In the following, we briefly discuss this challenge which, to the best of our knowledge, is not described in the OISC literature. For example, consider the *SUBLEQ* program in Listing 4.4 and assume that all *SUBLEQ* instructions operate on 3-bit values using two's complement: SRC1 holds the 3-bit value −4 and TMP0 holds the 3-bit value 0. In 3-bit two's complement, −4 and 3 are the smallest and largest possible values, respectively. Therefore, Line 1 leads to an erroneous branch since $0 - (-4) = 4$, which cannot be represented using 3-bits only, and thus the branch should not be taken. However, in 3-bit signed arithmetic using two's complements $0 - (-4) = -4$ which is $<= 0$ and thus the branch on Line 1, two instructions ahead is taken.

We solve this problem in ORISCV-ISS by computing the results using "enough" bits. Since the ISS is at the VP level, we cast the 32-bit values before the subtraction to 64-bit, perform the comparison on this 64-bit result, and then cast the result back to 32-bit before storing the result in a register. Therefore, our developed microcode shows the correct behavior. In a hardware implementation (using VHDL or Verilog), it is important that the OISC instruction can compute exact results in a mathematical sense.

4.4 Evaluation of ORISCV-ISS

The performance of a processor depends not only on the raw number of instructions it can execute in a given time but also, to a large extent, on other components such as the memory layout or the ISA. Since OISC architectures primarily target low-performance embedded devices, balancing between performance and chip area is of great importance. In this section, we demonstrate how ORISCV-ISS helps to find this balance.

We start from an ISS configuration in which only the *SUBLEQ* instruction is available, and all registers are mapped into the main memory. This represents our base case for maximum area efficiency, from which we explore possible improvements and their impact on the system performance (Sect. 4.4.1). To achieve this goal, we tune and explore two design characteristics: (1) the memory configuration and (2) extensions to the OISC ISA. First, we shift the memory-mapped registers toward dedicated registers for improved access times (Sect. 4.4.2). Then, we analyze the effect of an extended *SUBLEQ* architecture with specialized hardware for the most computationally intensive microcode procedures (Sect. 4.4.3).

4.4.1 Basic Timing Models

In this evaluation, we employed three basic memory configurations. Figure 4.2 visualizes these basic configurations as well as three additional configurations which we will describe in Sect. 4.4.2. The base assumption is that the execution of one *SUBLEQ* instruction takes one cycle.

Reading or writing to registers or the main memory adds a delay depending on the specific memory configuration. The first configuration, **RAM**, maps all RISC-V and OISC registers into the main memory. Since reading and writing to the main memory are slow, we add a delay of 4 additional cycles to each memory access. In the second configuration, **BRAM** all RISC-V and OISC registers are implemented

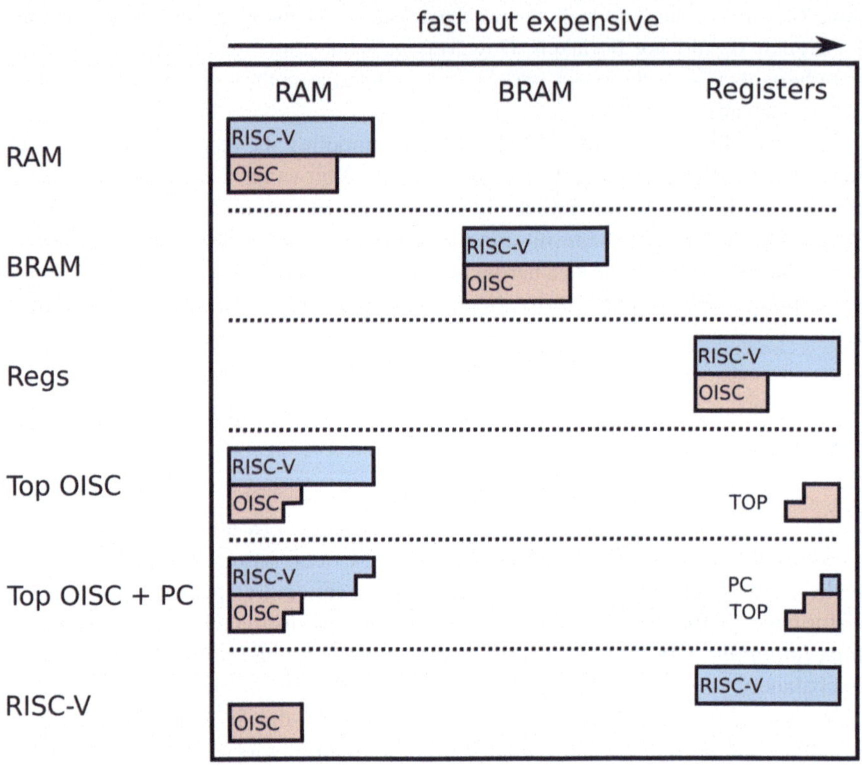

Fig. 4.2 Memory configurations for RISC-V and OISC registers

in larger on-chip memories like Block RAMs (BRAMs), but accessing the main memory for application data still requires 4 additional cycles. These memories are faster than the main memory and, thus, only add a delay of 1 for each read or write access. In the third configuration, **Regs**, we placed all RISC-V and OISC registers into dedicated registers. Reading and writing to these registers are fast and highly parallel, so no additional delay is added to the *SUBLEQ* execution time. The remaining three configs in Fig. 4.2 are described later when the context is clear.

4.4.2 Evaluation of Memory Configurations

In this section, we present a comparison of 6 memory configurations in terms of the number of cycles needed when executing software. We have taken several well-known benchmarks, compiled them to RV32I, and executed them on ORISCV-ISS in the respective memory configuration. Table 4.2 provides the results. The first column gives the name of the benchmark. The next three columns report the required

Table 4.2 Required microcode cycles for different memory configurations

	Basic Memory Configurations			Extra		
Benchmark	RAM	BRAM	Regs	Top OISC	Top OISC+PC	RISC-V
bit insert [9]	56,474	17,714	4,794	36,122	33,762	51,990
dijkstra [2]	753,261	241,062	70,329	378,093	311,569	620,341
heapsort [19]	96,21,348	3,100,233	926,528	4,836,100	4,029,276	7,990,048
bubblesort	$2.11 \cdot 10^8$	$6.8 \cdot 10^7$	$2.05 \cdot 10^7$	$9.49 \cdot 10^7$	$7.35 \cdot 10^7$	$1.67 \cdot 10^8$
ref [18]	$2.23 \cdot 10^8$	$7.26 \cdot 10^7$	$2.23 \cdot 10^7$	$1.1 \cdot 10^8$	$8.3 \cdot 10^7$	$1.73 \cdot 10^8$
ackermann [19]	$4.9 \cdot 10^8$	$1.58 \cdot 10^8$	$4.78 \cdot 10^7$	$2.53 \cdot 10^8$	$1.96 \cdot 10^8$	$3.84 \cdot 10^8$
sieve [19]	$4.29 \cdot 10^9$	$1.4 \cdot 10^9$	$4.33 \cdot 10^8$	$2.19 \cdot 10^9$	$1.67 \cdot 10^9$	$3.26 \cdot 10^9$
Avg. Reduction	Baseline	67.82%	90.43%	48.38%	58.00%	18.78%

microcode cycles according to the three *Basic Memory Configurations*. As can be seen in Table 4.2, implementing all registers in **BRAM** reduces the number of cycles on average by 67.82% w.r.t. the baseline. In comparison, shifting all RISC-V and OISC registers from the RAM to dedicated registers (**Regs**) cuts the required cycles more than 90%.

We now extended our design space exploration using ORISCV-ISS as follows: We wanted to find a configuration with minimal additional area costs but a significant cycle reduction. For this, we evaluated the **Top OISC** configuration, in which all registers are kept in the main memory except `SRC1`, `SRC2`, `TMP0`, and `TMP1` from the OISC unit. With just these four registers, we already observed on average a reduction of 48.38% w.r.t. the baseline. In configuration **Top OISC + PC**, also the RISC-V program counter is implemented in a dedicated register, since it is one of the most used RISC-V registers. By simply adding the RISC-V PC to the dedicated registers, we were able to additionally reduce the average number of cycles by nearly 10%. Finally, we evaluated a last configuration in which only the **RISC-V** registers are implemented in dedicated registers. Even though this configuration contains all 32 RISC-V registers and the RISC-V PC, the measured cycle reduction amounted to only 18.78%. This shows that the performance improvement highly depends on the selected registers. This was expected, since every RISC-V instruction requires multiple OISC instructions, and thus the OISC registers are used much more frequently, whereas the RISC-V registers are only used at the beginning and end of the execution of one RISC-V instruction.

4.4.3 Evaluation of SUBLEQ Extensions

Not every microcode procedure has the same complexity. For example, only 4 *SUBLEQ* instructions and cycles are required to execute the `addi` or `add` instructions. In contrast, all instructions that require bit-level access are much more complex. In our base microcode implementation, the `and` as well as `andi` microcode procedure

Table 4.3 Average cycles per RISC-V instruction

Benchmark	Base	BitOps
bit insert	24.5	12.3
dijkstra	12.9	12.3
heapsort	13.9	13.8
bubblesort	11.5	11.4
ref1	10.1	10.1
ackermann	11.3	11.3
sieve	10.0	10.0
Average	13.8	11.6

is 42 *SUBLEQ* instructions long. However, this procedure loops depending on the word width of the processor (in the case of RV32I, this is *XLEN*; 32) resulting in hundreds of microcode cycles for a single RISC-V instruction. This clearly shows that the performance depends on the to be executed instructions. However, this also creates opportunities to tailor the hardware to the specific application which we demonstrate in the following.

We measure how many microcode cycles it takes on average to execute one RISC-V instruction using the base microcode implementation. From there, we extend the ISS and the microcode using the functional registers to reduce the complexity of the most demanding microcode procedures.

Table 4.3 compares the average microcode cycles per RISC-V instruction for two microcode implementations. We can see that the **Base** implementation without any hardware support normally requires around 10 to 14 microcode cycles per RISC-V instruction. Of all benchmarks, only the *bit insert* benchmark makes heavy use of bit-level operations, whereas the other benchmarks mostly perform word-level arithmetic operations. This is clearly visible in the benchmark results, where the average cycles of the *bit insert* benchmark shoot up to 24.5 cycles per RISC-V instruction for the **Base** implementation.

To reduce the overhead for this benchmark, we added hardware support for the `and`, `or`, and `xor` RISC-V instructions[6] via the functional registers. The **BitOps** microcode implementation extends the **Base** configuration but replaces the costly bit operations with microcode procedures that target the functional registers. As a consequence, with this implementation, the number of cycles is now only half the number of cycles for the *bit insert* benchmark (first row in Table 4.3). The other benchmarks are mostly unaffected by the new microcode implementation, since they do not contain many calls to the bit-level instructions.

Additionally, the size of the **BitOps** microcode implementation with only 259 instructions is 25% smaller than the **Base** microcode implementation. This reduction is possible, since the complex procedures handling the bit operations in the base implementation are replaced with simple writes to the special registers.

[6] And also their immediate variants.

4.5 Formal Microcode Verification Framework

In Sect. 4.3.2, we presented the microcode layer of ORISCV-ISS. There, we also verified the correctness of our microcode using the RISC-V Architectural Tests which passed for all our 37 microcode procedures. Yet, writing correct SUBLEQ-based microcode is challenging, especially for RISC-V instructions which require many SUBLEQ instructions. Therefore, in this section, we present a formal microcode verification framework to ensure the correctness of our microcode procedures.

4.5.1 Initial SUBLEQ ISS

Our formal verification framework is based on a SUBLEQ ISS written in the Racket programming language. The SUBLEQ ISA is extremely minimal; therefore, the Racket ISS consists of just a single function (`step`, Listing 4.5) that totals about 30 lines of code which implements the SUBLEQ specification from Sect. 4.2. Please note that Listing 4.5 also contains some extensions to derive the formal model, which we describe in Sect. 4.5. These extensions are not part of the initial ISS and are highlighted using red color.

The `(step pc regs code)` function fetches one SUBLEQ instruction at position `pc` from the microcode in `code` (Line 4). Please note that Rackets `(list-ref xs i)` function returns the `i'th` element from list `xs`. Next, the *src1*, *src2*, and *jump* components of the instruction are extracted from the previously fetched SUBLEQ instruction (Lines 5–9). With the register addresses decoded, the values are read from the registers addressed by *src1* and *src2* (Lines 10–15). At this point, the values are sign extended to $XLEN + 1$ bits to overcome overflow errors which can impact the correctness of the SUBLEQ procedures [13]. The constant XLEN specifies the register bit-width of all registers and arithmetic operations. We verify RV32I instructions; therefore, XLEN is 32. However, we later show how XLEN can be adjusted to decrease the verification runtime of long-running SUBLEQ procedures. The main part of the SUBLEQ instruction, the subtraction, is implemented on Line 16. Next, the result is written back into the registers (Lines 18–23). The step function is implemented as a pure function without side effects. Therefore, a new list of registers `new-regs` is constructed which is identical to the list `regs` except in the place *src2* which holds the result of the subtraction. Racket's `for/list` function implements a *for each* loop with the addition that this function returns a list containing the results of all loop iterations. Subsequently, the next program counter is calculated on Lines 25–28. The `bvsle` function performs a signed less-or-equal operation on the bitvector arguments. Finally, the `step` function is called recursively with the previously computed new pc and register values (Line 31). However, if the end of the procedure is reached, which is indicated by the

EXIT constant in the jump offset (Line 30), the resulting register values in new-regs are returned (Line 32) and the SUBELQ execution stops.

As presented in Sect. 4.3.2.1, SUBLEQ instructions can access the 16 SUBLEQ registers shown in Table 4.1.

```
1  (define (step  fuel  pc regs code)
2    (cond [(= fuel 0) "unrolling depth too small"]
3      [else
4        (define instr (list-ref-bv code pc))
5        (define src1 (extract 15 12 instr))
6        (define src2 (extract 11 8 instr))
7        (define jump
8          (sign-extend (extract 7 0 instr)
9                       (bitvector 9)))
10       (define val1
11         (sign-extend (list-ref-bv regs src1)
12                      (bitvector (+ 1 XLEN))))
13       (define val2
14         (sign-extend (list-ref-bv regs src2)
15                      (bitvector (+ 1 XLEN))))
16       (define res (bvsub val2 val1))
17
18       (define new-regs
19         (for/list [(index (range 16))]
20           (if (bveq (integer->bitvector index (bitvector 4))
21                     src2)
22               (extract (- XLEN 1) 0 res)
23               (list-ref regs index))))
24
25       (define new-pc
26         (if (bvsle res (bv 0 (+ 1 XLEN)))
27             (bvadd pc jump)
28             (bvadd pc (bv 1 9))))
29
30       (if (bveq jump EXIT)
31           new-regs
32           (step (- fuel 1) new-pc new-regs code))]))
```

Listing 4.5 Step function of the verification ready model

4.5.2 *Formal SUBLEQ Execution Model*

This section describes the ISS extensions which we introduced to obtain the formal model. These extensions include the usage of symbolic input values, assumptions on the input values, and the inclusion of a maximum unroll depth of the model.

4.5.2.1 Symbolic Input

The first extension to the ISS concerns the input values, which are the initial register contents. While the registers contain concrete values in an ISS use case, they must contain symbolic values in a formal setting. Listing 4.6 showcases how symbolic registers can be created using Rosette's `define-symbolic` function. In our model, the register file is implemented as a list of registers. Therefore, the symbolic variables are created with the `define-symbolic` function (Lines 1–2) before they are inserted into a list (Lines 4–5).

```
1 (define-symbolic
2   val-tmp0 val-rvpc ... val-tmp7 (bitvector XLEN))
3
4 (define init-regs
5   (list val-tmp0 val-rvpc ... val-tmp7))
```

Listing 4.6 Initialization of the symbolic registers

4.5.2.2 Constant Assumption

Since all registers are symbolic variables, we must make sure that the variables for the constants (e.g., `INC` or `NEXT`) always contain the correct values. Therefore, we use the Rosette's `assume` function to force the register values to take the constant values at the very start of the model before the first call to *step*. The `assume` function reduces the search space by dropping all variable assignments that violate the assumptions. Listing 4.7 shows the call to Rosette's `assume` function to initialize the correct constants in the initial register values. `REG-ONE`, `REG-WORD`, `REG-INC`, and `REG-NEXT` are definitions of the internal register numbers for the registers `ONE`, `WORD`, `INC`, and `NEXT`, respectively. The assumption condition is a conjunction of multiple equivalence checks (via Racket's `eq?` comparison function) each testing if the register holds the correct value. Since we use Rosette on problems with bounded arithmetic, the values are specified using the `(bv v w)` function. This function constructs a bitvector representation of value `v` using `w` bits.

```
(assume
 (and
  (eq? (list-ref init-regs REG-ONE)  (bv 1 XLEN))
  (eq? (list-ref init-regs REG-WORD) (bv (- XLEN 1) XLEN))
  (eq? (list-ref init-regs REG-INC)  (bv -1 XLEN))
  (eq? (list-ref init-regs REG-NEXT) (bv -4 XLEN))))
```

Listing 4.7 The constant registers are set using the Rosettes assume function

4.5.2.3 Unroll Depth

During formal verification, the model is unrolled depending on the number of executed SUBLEQ instructions. Rosette does not support unbounded loops and unbounded recursion. Therefore, we have to explicitly specify the maximum number of unrollings. This unroll depth is one of the major factors for the verification runtime. Therefore, minimizing the unroll depth is of utmost importance. Since we know the worst case run times of all SUBLEQ procedures, an upper bound for the unroll depth can be determined. We integrate this unroll depth into the model by providing an additional argument *fuel* (Line 1 in Listing 4.5). In other words, this argument specifies how many SUBLEQ instructions can still be executed until the result of the specified RISC-V instruction is placed in the RSLT register. At each recursive call to `step`, the fuel argument is decremented. If the initial fuel parameter was chosen too small, the procedure terminates with a respective error message (Line 2 in Listing 4.5), since the SUBLEQ microcode procedure did not meet the specification of the RISC-V instruction at hand. Now, the model is ready to be used for verification with Rosette. The last remaining piece to a full verification setup is the RISC-V specifications, which we will describe in the following section.

4.5.3 Definition of RISC-V Specifications

To formally capture the specification of each RISC-V instruction, we have defined the macro `rv-verify`. This macro generates Racket code which handles the formal verification, verification runtime measurement, and the concrete execution of models, if a model is found. Listing 4.8 shows how the `rv-verify` macro is used for the *JAL* RISC-V instruction. Before we discuss the concrete definitions for *JAL*, we explain the general meaning of the 7 parameters used in the macro and the general task of the macro in the following.

4.5.3.1 `rv-verify` Parameters

Name	The name of the RISC-V instruction which is implemented by the SUBLEQ microcode procedure.
PC	The microcode program counter for fetching the first SUBLEQ instruction of the RISC-V instruction at hand.
Fuel	The maximum number of unrollings.
Microcode	The microcode containing all SUBLEQ instructions.
Solver	The SMT solver used during verification.
Spec	The specification of the RISC-V instruction which defines the functional behavior that must hold on all valid inputs. For branching operations, the only specification is the correct adaption of the program counter (pc+offset in case of a jump, pc + 4 otherwise).

For other instructions, it ensures that the result of a calculation is correct and stored in the right register. The registers for pc and the result are the SUBLEQ registers RVPC and RSLT, respectively. For both, the position in the memory is known. This means that we can access their content via the respective named index (see Lines 9 and 11).

Assumptions: Assumptions that restrict valid inputs for some instructions. For example, the shift amount for shifting instructions is limited to 5 bits (≤ 31).

4.5.3.2 `rv-verify` General Task

Our `rv-verify` macro is responsible for executing the respective SUBLEQ instruction, printing the result after execution, verifying that the specification holds, and providing information about the runtime of the verification. The specification and assumptions have to be passed in a lambda expression, as they would otherwise be evaluated immediately (Lines 8–16). This immediate evaluation is not possible as both refer to register values *after* execution which are not known to the macro at the time of definition. The macro first symbolically executes the instructions starting at the defined program counter and saves the resulting state of all registers. The selected SMT solver then tries to find concrete values for the defined symbolic constants, which would lead to a violation of the specification. If no such counterexample is found, the correctness is proven, and the function finally returns OK and the runtime for the verification. Otherwise, it returns a model, which consists of the register values that led to a wrong result after executing the instructions and prints the incorrect output.

4.5.3.3 `rv-verify` for *JAL*

Listing 4.8 shows the specification for SUBLEQ procedure of the RISC-V instruction *JAL*. As it is a J-Type instruction, the program counter needs to be adjusted by the offset to the instruction that should be executed next (RVPC =RVPC+IMM), and the previous program counter gets incremented by 4 and written into the destination register RSLT. For this specification, we assume on Line 13 that the *JAL* instruction gets a valid value for the program count, namely, 2-bit aligned (lowest two bits are 0). All registers contain bitvectors; therefore, the behavior of the addition is defined with `bvadd`, one of Rosette's bitvector operations. The equality of the content of the RSLT register and the desired result is checked using the `eq?` predicate, which returns true if two values or their contents are equal.

```
1 (rv-verify
2  #:name "JAL"
3  #:init-pc JAL-PC
```

```
#:fuel 20
#:microcode microcode
#:solver (boolector)
#:spec
  (lambda (res)
    (and (eq? (list-ref res REG-RVPC)
              (bvadd val-rvpc val-immi))
         (eq? (list-ref res REG-RSLT)
              (bvadd val-rvpc (bv 4 XLEN)))))
#:assumptions
  (lambda (res)
    (assume (eq? (bv 0 2)
                 (extract 1 0 (list-ref res 1))))))
```

Listing 4.8 Full specification of the JAL instruction

4.5.4 Optimization Techniques

A large part of the RV32I instructions can be formally verified with our framework. By using a formal approach, we were able to identify microcode bugs which were not uncovered by testing and the RISC-V architectural test suite. However, some procedures showed runtimes of much more than 12 hours, which we set as the time limit. For these procedures, we had to introduce special optimizations to make them complete in under 12 hours.

4.5.4.1 Microcode Optimization

The procedures with long runtimes are exclusively procedures that have internal loops that lead to very deep unroll depths. Especially the number of instructions inside the internal loops of microcode procedures contributes to the required unroll depth. Minimizing the number of instructions inside those internal loops is especially effective, since they are multiplied by the XLEN, and thus even one instruction fewer reduces the unroll depth by up to 32. Therefore, as a first measure to reducing the verification runtime, we spend significant effort in optimizing the procedures. For some instructions, we were able to remove more than ten instructions which, assuming $XLEN = 32$, reduces the required unroll-depth already by ≥ 320.

4.5.4.2 Register Bit-Width

All procedures that cannot be verified given our time limit share the property that their unrolling depth depends on the bit-width of the registers. For example, logical operations such as *XOR* have to perform multiple SUBLEQ instructions in a loop for each register bit. Therefore, decreasing the bit-width (XLEN) of the formal model directly reduces the required unroll depth. By this, we were able to verify the *AND*, *OR*, and *XOR* instructions and their immediate counterparts.

We argue that in the case of these logical instructions, all bits are independent. Therefore, instead of proving correctness for 32 bits, we can safely scale down the size of the bitvectors.

4.6 Microcode Verification Results

In this section, we present the results for formal verification of SUBLEQ procedures implementing RV32I. Our experiments have been carried out on an Intel Core i7-10700 with 64 GB of main memory. For SMT-solving, we used Boolector [20] with the timeout set to 12 hours (43,200 seconds).

In the following, we evaluate the verification runtime of all microcode procedures (Sect. 4.6.1), describe one found bug in detail (Sect. 4.6.2), and analyze how the verification runtime scales with increasing word lengths (Sect. 4.6.3).

4.6.1 Overall Results

Table 4.4 shows the verification results for all RISC-V instructions. Column *Instruction* provides the name of the RISC-V instruction that is implemented by the corresponding SUBLEQ procedure. Please note that one SUBLEQ procedure can implement multiple RISC-V instructions, e.g., the microcode for ADD and ADDI is the same as the only difference is that the second source is either the RISC-V register rs2 or the immediate `imm` which is handled by the RISC-V Interface by making the respective value in both cases available in the SUBLEQ register `IMM`. Column *#SUBLEQ Instrs* gives the number of SUBLEQ instructions needed to implement the RISC-V instruction. Then, Column *Unrolling depth* reports how often we had to unroll the `step` function for the proof.

Some SUBLEQ procedures are loop-free (e.g., ADDI or BLT); therefore, the model has to be unrolled exactly as many times as the number of SUBELQ instructions in the corresponding microcode procedure. Other instructions are more complex and must be implemented using loops in the SUBLEQ procedure (e.g., XORI or SLL). For these procedures, the number of unrollings must be individually calculated. For example, the *XORI* instruction consists of 7 initial instructions that are always executed just once, followed by a loop consisting of 18 instructions. The loop is executed for each bit of the arguments (i.e., XLEN times). After the loop, a single instruction which increments the RISC-V program counter is always executed. Overall, the maximum unroll depth for $XLEN = 32$ is therefore $8 + 18 * XLEN = 8 * 18 * 32 = 584$.

Table 4.4 Verification results for all RISC-V instructions

Instruction	#SUBLEQ Instrs	Unroll depth	Result	Runtime (s)
LB	4	4	Pass	0.02
LH	4	4	Pass	0.02
LW	4	4	Pass	0.02
LBU	4	4	Pass	0.02
LHU	4	4	Pass	0.02
SB	4	4	Pass	0.02
SH	4	4	Pass	0.02
SW	4	4	Pass	0.02
LUI	4	4	Pass	0.02
ADD	4	4	Pass	0.02
ADDI	4	4	Pass	0.02
SUB	6	6	Pass	0.04
AUIPC	8	8	Pass	0.02
JAL	8	8	Pass	0.03
JALR	16	16	Fail	0.03
BEQ	11	11	Pass	2.76
BNE	11	11	Pass	2.30
BLT	6	6	Pass	3.79
BGE	5	5	Pass	4.35
BLTU	20	485	Fail	1,163.00
BGEU	28	617	Fail	1.26
SLT	6	6	Pass	2.07
SLTI	6	6	Pass	2.07
SLTU	29	618	Fail	165.26
SLTIU	29	618	Fail	165.26
XOR*	26	314	Pass	15,183.00
XORI*	26	314	Pass	15,183.00
OR*	25	297	Pass	13,107.00
ORI*	25	297	Pass	13,107.00
AND†	24	234	Pass	41,352.00
ANDI†	24	234	Pass	41,352.00
SLL	8	132	Pass	280.00
SLLI	8	132	Pass	280.00
SRL	23	550	Fail	14.84
SRLI	23	550	Fail	14.84
SRA	28	555	Fail	14.84
SRAI	28	555	Fail	14.84

* XLEN=17 † XLEN=15

The verification result is provided in Column *Result*, and the runtime in seconds is reported in Column *Runtime*. Finally, instructions for which we had to scale down the size of the bitvectors to get practical runtimes are marked by * and †. For example, since operations on single bits (e.g., logical and) are challenging to implement using SUBLEQ instructions only, we had to reduce the bit-width for the *AND(I)*, *OR(I)*, and *XOR(I)* procedures. This reduced the required depth for unrolling significantly.

Overall, as can be seen, we were able to prove the correctness of SUBLEQ microcode procedures for the majority of the RV32I instructions (28 out of 37). However, for 9 RISC-V instructions, the corresponding SUBLEQ procedures still contained bugs. After discovering the 9 faulty microcode procedures, we could fix all of them. Debugging the faulty procedures was made easier by the counterexamples produced by the formal model.

4.6.2 *Example Bug*

One of the bugs we found using our framework concerns the correctness of the *JALR* instruction. This instruction adds an immediate to the source register, then has to clear the LSB of the sum, and then jumps to the resulting address. However, our SUBLEQ microcode procedure for *JALR* does not set the LSB to zero, and thus invalid program counter values can be reached. The problem occurs when the sum of the source register and the immediate value is odd, which is possible, since *JALR* uses the I immediate type and the source register can also hold odd values.

4.6.3 *Effect of Word Lengths*

Figure 4.3 shows the runtimes for verifying the SUBLEQ procedures implementing *AND(I)*, *OR(I)*, *XOR(I)*, and *SLL* for various word lengths (i.e., XLEN values). As can be seen, *AND(I)*, *OR(I)*, and *XOR(I)* show exponential runtimes with increasing XLEN. Starting from around $XLEN = 10$, the runtime is increasing significantly for these two microcode procedures with *AND(I)* reaching runtimes of multiple hours from $XLEN = 13$ on. *SLL* on the other hand scales much better, with the full $XLEN = 32$ verification taking only 237 seconds. Please note the runtime differences between the logical and instruction *AND(I)* and the other two instructions *XOR(I)* and *OR(I)*. Despite nearly identical unroll depths, the runtime of *AND(I)* scales much worse than *XOR(I)* and *OR(I)*.

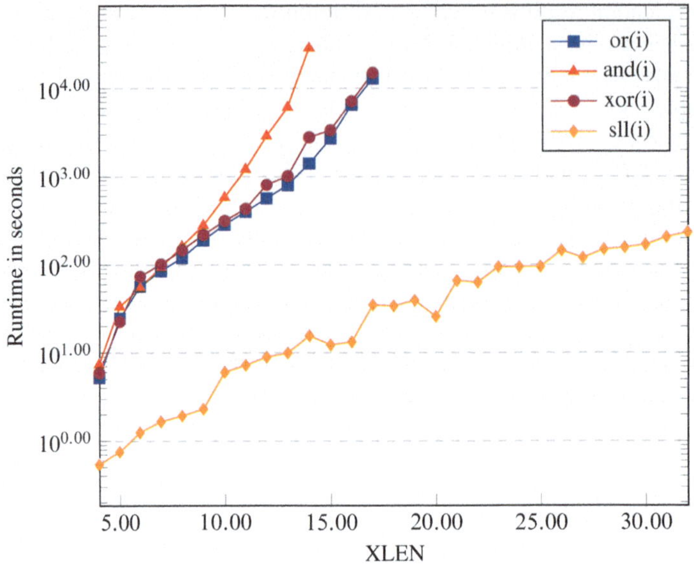

Fig. 4.3 Verification runtime with increasing XLEN

4.7 Summary

In this chapter, we presented a microcoded and formally verified RISC-V platform based on an OISC architecture that implements the RV32I ISA. All instructions of the RV32I ISA are implemented in microcode which makes use of only one instruction, SUBLEQ. We developed the RISC-V platform in two steps. First, we presented ORISCV-ISS, a VP we used to quickly develop the SUBLEQ microcode and for design space exploration. We tested the compliance of ORISCV-ISS and our microcode using the RISC-V architectural tests, which passed for all instructions.

Next, we presented a verification framework for our SUBLEQ microcode procedures implementing the RV32I ISA. We created our formal model for SUBLEQ procedures in Rosette, a solver-aided programming language. This allowed us to extend an already existing ISS for SUBLEQ procedures with constraints capturing (a) the execution of a single SUBLEQ instruction symbolically and (b) the formalization of the RISC-V instruction specifications in terms of lambda-based macros. Moreover, we presented two techniques to tackle the verification complexity of long-running microcode procedures. Our methodology uncovered multiple microcode bugs which were not found by official RISC-V architectural tests. All the uncovered bugs were corrected, and the verification is now successful for all microcode procedures.

In future work, we plan to investigate different SMT solvers as well as advanced techniques such as predicate abstraction. Moreover, we are working on an FPGA core which implements the RISC-V interface and runs the verified SUBLEQ microcode. For the full verification of the core, we will leverage novel approaches such as EPEX, which we presented in the previous chapter. Additionally, we will analyze different metrics using the waveform analysis language, which we present in the next chapter.

References

1. RISC-V architecture test sig. https://github.com/riscv-non-isa/riscv-arch-test. Accessed: 2024-04-18
2. R. Abishek, Dijkstra algorithm implementation. https://gist.github.com/rajabishek/2c75052a674fd15e2170. Accessed: 2024-04-18
3. S. Ananthanarayan, S. Garg, H.D. Patel, Low cost permanent fault detection using ultra-reduced instruction set co-processors, in *Design, Automation and Test in Europe* (2013), pp. 933–938
4. O. Bringmann, W. Ecker, A. Gerstlauer, A. Goyal, D. Mueller-Gritschneder, P. Sasidharan, S. Singh, The next generation of virtual prototyping: ultra-fast yet accurate simulation of hw/sw systems, in *Design, Automation and Test in Europe* (2015), pp. 1698–1707
5. A. Charif, G. Busnot, R. Mameesh, T. Sassolas, N. Ventroux, Fast virtual prototyping for embedded computing systems design and exploration, in *RAPIDO Workshop* (2019), pp. 3:1–3:8
6. M. Crepaldi, A. Merello, M. Di Salvo, A multi-one instruction set computer for microcontroller applications. IEEE Access **9**, 113454–113474 (2021)
7. T. De Schutter, *Better Software. Faster!: Best Practices in Virtual Prototyping* (Synopsys Press, California, 2014)
8. K.S. Dharshana, K. Balasubramanian, M. Arun, Encrypted computation on a one instruction set architecture, in *International Conference on Circuit, Power and Computing Technologies* (2016), pp. 1–6
9. Hansraj Das. Algorithm repository. https://github.com/hansrajdas/algorithms. Accessed: 2024-04-18
10. V. Herdt, D. Große, R. Drechsler, *Enhanced Virtual Prototyping: Featuring RISC-V Case Studies* (Springer, Berlin, 2020)
11. P. Jääskeläinen, A. Tervo, G.P. Vayá, T. Viitanen, N. Behmann, J. Takala, H. Blume, Transport-triggered soft cores, in *IEEE International Symposium on Parallel and Distributed Processing Workshops and Phd Forum* (2018), pp. 83–90
12. D.W. Jones, The ultimate RISC. SIGARCH Comput. Architecture News **16**(3), 48–55 (1988)
13. L. Klemmer, D. Große, An exploration platform for microcoded RISC-V cores leveraging the one instruction set computer principle, in *IEEE Annual Symposium on VLSI* (2022), pp. 38–43
14. L. Klemmer, S. Gurtner, D. Große, Formal verification of SUBLEQ microcode implementing the RV32I ISA, in *Forum on Specification and Design Languages* (2022), pp. 1–8
15. F. Mavaddat, B. Parhami, URISC: The ultimate reduced instruction set computer. Int. J. Electr. Eng. Educ. **25**(4), 327–334 (1988)
16. Maxim Integrated. Maxq microcontroller family. https://www.maximintegrated.com/en/design/technical-documents/userguides-and-manuals/5/5618.html. Accessed: 2024-04-18
17. O. Mazonka, Bit copying—the ultimate computational simplicity, in *arXiv:0907.2173.2593* (2009)
18. O. Mazonka, A. Kolodin, A simple multi-processor computer based on subleq, in *arXiv:1106.2593* (2011)
19. Microsoft. Microsoft test-suite. https://github.com/microsoft/test-suite. Accessed: 2024-04-18
20. A. Niemetz, M. Preiner, C. Wolf, A. Biere, Btor2, BtorMC and Boolector 3.0, in *Computer Aided Verification* (2018), pp. 587–595

21. A. Rajendiran, S. Ananthanarayanan, H.D. Patel, M.V. Tripunitara, S. Garg, Reliable computing with ultra-reduced instruction set co-processors, in *Design Automation Conference* (2012), pp. 697–702
22. M.M. Shulaker, G. Hills, N. Patil, H. Wei, H.-Y. Chen, H.-S.P. Wong, S. Mitra, Carbon nanotube computer. Nature **501**(7468), 526–530 (2013)
23. M. Yokota, K. Saso, Y. Hara-Azumi, One-instruction set computer-based multicore processors for energy-efficient streaming data processing, in *IEEE International Workshop on Rapid System Prototyping* (2017), pp. 71–77

Chapter 5
The Waveform Analysis Language

This chapter introduces the Waveform Analysis Language (WAL). Waveforms are a central data format that is relevant at most stages of the hardware development process from simulation, over (formal) verification, to debugging via logic analyzers. They contain large amounts of valuable information, for example, about the performance or correctness of the design. However, extracting this information is challenging, not least due to the often complex analysis logic required to do so and the large amounts of data in waveforms. WAL is a Domain Specific Language (DSL) for efficiently and automatically performing analysis tasks such as verification or performance analysis on waveforms. The following sections present the language, design decisions, and several case studies.

5.1 Introduction

The development of next-generation electronic systems poses significant challenges to all phases of the design process. In particular, the verification phase is the most time-consuming part. Verification aims to find design errors as early as possible, and verification is multidimensional [11]: Both functional requirements and nonfunctional requirements (e.g., timing, performance, latency) have to be verified, and in practice, there is often an intersection of the respective tasks and models used [4]. If one again breaks down the verification task, then it is dominated by debugging [7]. Even worse, debugging is rated as least predictable, since it requires a deep design understanding [1]. The cornerstone for design understanding is the waveform. Waveforms are generated during simulation of HDL designs and describe the circuit signals together with hierarchy information over time [2].

In both the design phase and the verification phase of a digital system, waveforms are heavily used. Initially, directed test stimuli are created to see that the currently designed Hardware (HW) blocks are "alive" and produce some meaningful output.

L. Klemmer, D. Große, *Versatile Hardware Analysis Techniques*,
https://doi.org/10.1007/978-3-031-83093-8_5

Once the design matures, the verification plan is followed, and advanced verification techniques, e.g., assertion-based methods together with coverage-based solutions, are employed [8, 22]. Along this highly iterative process, waveforms demonstrating expected behavior or unexpected behavior (e.g., in case of a failed assertion or a violated timing constraint) have to be analyzed and understood. For this task, (commercial) waveform viewers are utilized. Waveform viewers are software tools which allow viewing signal values over time. Besides selecting the radix of each signal and grouping signals together, the user can zoom in and out, jump to the next time point where the value of a signal changes, determine the time difference between two cursors, etc. However, while all these features help in understanding and debugging, *waveform viewing is a highly manual and tedious process*.

So far, most research has either concentrated on specific design understanding approaches such that the manual analysis of the waveform is reduced to a minimum, or it has been confined on the generation of "better" waveforms, e.g., by employing formal methods, reducing their length, or minimizing the signals involved in a failing trace. While these approaches have introduced automation in general, there has been almost *no progress for automating the analysis of waveforms*. The potential for automation becomes clear when looking at typical nontrivial analysis questions raised for waveforms dumped, for instance, when running Software (SW) on a processor:

- How many instructions per cycle are executed by the processor?
- What is the latency of the bus interfaces?
- What throughput is the bus achieving?
- When and why is the processor pipeline flushed or stalled during software execution?
- Which parts of a program are executed on the processor?

In this chapter, we *bring automation to the analysis of waveforms* and introduce the Waveform Analysis Language (WAL).[1] We have implemented WAL as a DSL [9]. WAL is based on the idea that *signals from waveforms become variables* and that *simulation time and design hierarchy are fundamental parts* of the language.

WAL Basis For WAL, we first identified the essential operations for processing waveform data; this includes loading (multiple) waveforms, access to signals, time manipulation, and logical grouping of signals. Second, since verification (and regression) environments differ widely in terms of requirements and workflows, we strive to maximize the versatility of WAL. Therefore, we decided to define the syntax of our proposed WAL DSL following the established Lisp principle of *symbolic expressions* (or S-expressions) [21]. S-expressions have three main advantages: an extremely regular syntax based on lists, code and data are represented by the same

[1] The version of WAL used in this book is open-source and available at https://github.com/ics-jku/wal/tree/f6e31721f147e3131a3a01e4bee03ff63b8b0bb3.

data structure, and code generation is simple as only lists have to be created. As a consequence, we can:

1. feature a minimal and clean syntax for WAL,
2. easily integrate the abovementioned essential waveform operations as well as advanced operations via functions,
3. provide an *Intermediate Representation* (IR) well-suited as a compilation target from other languages, and
4. introduce a macro system to WAL which makes extending and modifying the language simple, opening the door to additional DSLs build in WAL itself.

To demonstrate the capabilities of WAL, we present four different case studies. The case studies cover the combined HW/SW profiling, bus transaction analysis, processor pipeline analysis, and the implementation of an embedded DSL using WALs macro system.

This chapter includes and extends published material from one journal article [17], three conference papers [14, 15, 18], and one workshop paper [16]. We begin this chapter by discussing the waveform analysis problem in Sect. 5.2. Then, we introduce WAL in Sect. 5.3. There we present the requirements, review S-expressions in the context of WAL, and introduce the essential and advanced operations of WAL including the macro system and new time handling functionality. Further, we discuss WAL core, our reference implementation of WAL in Sect. 5.4, and evaluate WAL on several case studies in Sect. 5.5. Finally, we summarize this chapter and discuss WAL's current limitations and possible future work in Sect. 5.6.

5.2 Programmable Waveform Analysis Problem

Design understanding and debugging is an active field of research. Several methods targeting specific problems and different abstraction levels have been proposed so far [26]. For example, this includes natural language techniques to derive assertions from specifications [37], feature localization in ESL models [23] or in RTL descriptions [20], assertion mining at RTL [32], identification of instruction pipelines using static analysis on the netlist [28], template-based understanding of circuit components [10], and reverse engineering at the gate-level [19, 31]. However, all these solutions focus on dedicated design understanding of subproblems and do not provide a generic user-programmable analysis for waveforms.

In general, advanced testbench constructs, for example, provided by SystemVerilog, are relevant in this context. SystemVerilog offers several features and methodologies that can help in monitoring, capturing, and analyzing signal behaviors during simulation. This includes assertions, as already mentioned in the introduction. They capture complex temporal behaviors and are used to flag unexpected signal values or sequences during simulation.

However, the goal of assertion-based verification is to determine "only" whether the underlying temporal logic formula evaluates to true or false on a trace (or waveform). In contrast, WAL allows a much wider user-programmable analysis and application: With WAL programs, complex signal relations can be caught as well, but then the user can perform arbitrary actions for all kinds of computations including the use of high-level data structures, such as lists, hash maps, etc. In principle, a user could perform certain analysis in the testbench of a design by utilizing, for instance, classes and the object-oriented features of SystemVerilog. However, such an approach is extremely complex and would require significant effort and also resimulation, quickly reaching the limits of practicability.

The *UVM* [13] does not directly provide features for programmable waveform analysis, as it is primarily a methodology and a library for creating structured, reusable verification environments in SystemVerilog. UVM focuses on the creation, management, and the use of verification components and data, such as stimulus generation, checking mechanisms, coverage collection, and reporting. Its primary goal is to improve the efficiency and reusability of the verification process.

To the best of our knowledge, no waveform analysis language comparable to the user-programmable expressiveness of WAL is currently available (also not in the commercial solutions from Cadence, Synopsys, Siemens EDA, etc.).

As already stated in the introduction, after the generation of a waveform from a HW simulation (typically in a format such as Value Change Dump (VCD) [12]), the contained signals and their relations are traditionally analyzed by visual inspection using waveform viewers. The major reason is that waveforms are an elegant method of visually expressing concurrency and design hierarchy. Waveforms typically contain not only the value of signals over the simulation time but also information about the signals (bitwidth, type) and about hierarchical scope information, i.e., which modules make up the design and how they are composed. Hence, debugging and understanding of sophisticated module behavior and inter-module interactions can be performed. However, as the analysis is manual, it can quickly become tedious or totally impractical in the case of complex or repeating problems. Therefore, we introduce user-programmable waveform analysis, transforming a manual repeating analysis problem into a one-time-only effort that scales with increasing complexity and waveform sizes and which is often reusable across projects.

In the following, we provide a simple but illustrative example for such an analysis problem. We will use this example throughout the chapter when introducing the proposed WAL. Figure 5.1 shows the waveform of a bus communication using the typical request-acknowledge protocol scheme.

Two components (*comp1* and *comp2*) are connected to the bus, and the respective *req*/*ack* signals have been traced as can be seen in the waveform. Assume that the design team has to determine the average latency for each component attached to the bus for system optimization. To solve such a problem with a waveform viewer is practically not possible. More precisely, the task is to walk through the trace and count cycles where *req* is high and *ack* is low. Finally, this result has to be divided by the number of acknowledgments. In Fig. 5.1, for *comp1*, we get $3 + 2 = 5$ (see marked lines a to b and c to d, respectively) divided by 2, which gives the average latency of 2.5. Clearly, manual navigation and calculation on the waveform is a poor solution only. Even worse, for waveforms with tens of thousands of cycles,

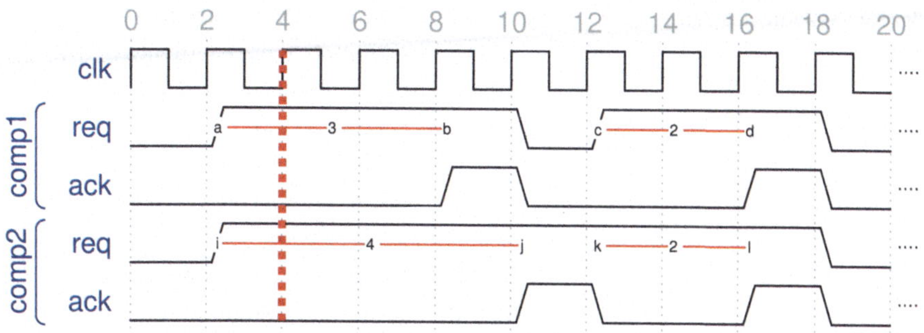

Fig. 5.1 Request-acknowledge bus communication

this approach fails. In contrast, this calculation can be easily performed with WAL as we will show later.

Moreover, the design in the example has two components, and therefore we are interested in the average latency over all, or a subset, of the components. This requires to extend the calculation from *comp1* to *comp2*, which would lead to doubling the code. For such problems, we added advanced features to WAL which allow for writing code in a generic and flexible way.

In the next section, we introduce WAL and show that the average latency can be easily determined using WAL.

5.3 Waveform Analysis Language (WAL)

First, we consider the requirements on the design and the implementation of the WAL DSL (Sect. 5.3.1). Then, we briefly review symbolic expressions as proposed by Lisp, which we extend to accommodate the specifics of WAL and the HW domain (Sect. 5.3.2). We continue with the essential WAL operations (Sect. 5.3.3) and WALs macro system (Sect. 5.3.4). Finally, we close this section with a presentation of the advanced WAL operations (Sect. 5.3.5).

5.3.1 Design and Implementation Requirements

After conceptualizing WALs functional scope and before starting a first implementation, we had to decide upon a suitable architecture. At first glance, it seems advantageous to implement all functionality in a library of an established programming language (i.e., Python or C) as this provides a proven basis and is easy to pick up for most developers. Unfortunately, this approach has significant drawbacks to the versatility and expressiveness of WAL as all feasible languages are geared

Fig. 5.2 Simplified BNF of WAL

```
⟨expression⟩ ⊨ ⟨atom⟩ | ⟨list⟩ | ⟨quoted⟩ |
               ⟨sliced⟩ | ⟨timed⟩ | ⟨resolved⟩ | λ
      ⟨atom⟩ ⊨ number | symbol | string
    ⟨quoted⟩ ⊨ '⟨expression⟩ |
               `⟨expression⟩ |
               ,⟨expression⟩ |
               ,@⟨expression⟩
  ⟨resolved⟩ ⊨ #⟨expression⟩ | ~⟨expression⟩
    ⟨sliced⟩ ⊨ ⟨expression⟩[⟨expression⟩]
               ⟨expression⟩[⟨expression⟩:⟨expression⟩]
     ⟨timed⟩ ⊨ ⟨expression⟩@⟨expression⟩
      ⟨list⟩ ⊨ (⟨expression⟩⟨element⟩)
               [⟨expression⟩⟨element⟩]
               {⟨expression⟩⟨element⟩}
   ⟨element⟩ ⊨ whitespace⟨expression⟩⟨element⟩ | λ
```

toward general purpose computing. This means that expressing waveform analysis problems would require large amounts of boilerplate code, for example, "getter" functions for signal access, as many waveform-specific actions are not native to the language. However, we envisioned a system where all aspects of the waveform and HW design domain are **first-class citizens of the language**. Designing WAL as a DSL with waveform analysis in mind enables users to directly express their problems naturally instead of forcing their problem onto the paradigm of a different language. Finally, even though we developed a reference implementation of WAL from the ground up, it is possible to implement WAL on top of other programming languages (similar to languages implemented on top of Racket [25]).

5.3.2 S-Expression and WAL-Specific Syntax

Symbolic expressions (abbrev. as S-expressions) are common in languages related to Lisp, such as Common Lisp [5] or Scheme [24]. Fundamentally, S-expressions can be of two kinds: *atoms* or *lists*. Atoms are literals like numerical values (e.g., `1`, `0xff`), string values (e.g., `"text"`), booleans (e.g., `true`, `false`), or *symbols*. Lists are multiple S-expressions separated by white space and enclosed in parentheses `(expr1 expr2 ...)`. All operators and function calls are written in prefix notation, e.g., `(+ 3 b)` to compute the sum of 3 and b.

A simplified definition of WALs syntax is shown in Backus-Naur Form (BNF) form in Fig. 5.2. Due to space limitations, the definitions for numbers, symbols, and strings are omitted (however, they mostly follow SystemVerilog formats). In particular, Fig. 5.2 focuses on the WAL-specific additions to the S-expression syntax, namely, the *resolved*, *sliced*, and *timed* constructs. These three special

Table 5.1 Special shorthand syntax in WAL

Special syntax	Transformed into
`'expr`	`(quote expr)`
`` `expr ``	`(quasiquote expr)`
`,expr`	`(unquote expr)`
`expr@off`	`(reval expr off)`
`~symbol`	`(resolve-scope symbol)`
`#symbol`	`(resolve-group symbol)`
`expr[i]`	`(slice expr i)`
`expr[h:l]`	`(slice expr h l)`

language features are valid WAL syntax. However, they are transformed into regular S-expressions as shown in Table 5.1.

Now, let us look at S-expressions in WAL; i.e., we consider them in the context of waveforms. As a consequence, the *symbols* of S-expressions are either signal names contained in a waveform, e.g., `top.module1.out`, or variable names defined in a WAL program. With respect to evaluation of an S-expression, we define the *current time index* (or just index) for a waveform at hand. In Fig. 5.1, this is nothing else than the red dashed line shown at position 4. So a signal name (symbol) is evaluated to the value of the signal at the current time index, for instance, `comp1.req = 1` and `comp1.ack = 0` at the current time index 4.

Besides the access to signal values, all operations targeting the analysis of waveforms are integrated into WAL S-expressions using dedicated functions. In the following sections, we introduce these functions and demonstrate how they allow formulating compact and easy-to-use programs for waveform analysis.

5.3.3 *Essential WAL Operations*

We provide an intuitive introduction to the essential operations of WAL, and therefore we incrementally develop a WAL program to solve the latency analysis problem as presented in Sect. 5.2. For the essential WAL operations, three main categories can be distinguished: waveform handling, signal access, and timing. As a foundation for all WAL expressions, WAL naturally implements all basic programming constructs (e.g., variables, loops, user functions). Table 5.2 (starting on Page 66) summarizes most WAL functions, and in the following, we always refer to this table.

5.3.3.1 Waveform Handling

First, a waveform must be loaded in a WAL program in order to have access to the signal values. The `load` operator reads the waveform specified by the first argument and registers it with the optional *id* given as the second argument. Assume

Table 5.2 Overview of WAL functions

Name	Description	Example
`load`	Loads trace into WAL, with optional trace id `t`	`(load "trace.fst" 't)`
`unload`	Removes trace with trace id `t` from WAL	`(unload 't)`
`step`	Adds n to INDEX	`(step n)`
`eval-file`	Evaluates file `f` and makes definitions available	`(eval-file f)`
`alias`	Creates an alias for a signal name	`(alias short "very.long.signal.name")`
`unalias`	Removes an alias	`(unalias short)`
`get`	Gets value from signal specified by string or symbol	`(get "top.sub.ready")`
`reval`	Locally modifies INDEX and evaluates body expression	`(= top.signal (reval top.signal -1))`
`import`	Imports external functions (e.g., from Python)	`(import riscv)`
`call`	Calls function from imported file	`(call riscv.decode tb.top.instr)`
`+,-,*,/,&&,||`	Arithmetic and logical functions	`(* a b (+ 1 2))`
`<,>,=,<=,>=`	Comparison functions	`(< 1 2)`
`define`	Defines `x` and bind value `v` to it	`(define x v)`
`set!`	Sets variable `x` to value `v`	`(set! x v)`
`let`	Binds value `v` to `x` and evaluate body	`(let ([x v]) (+ x 4))`
`if`	Conditional Branch	`(if (< x 5) "small" "large")`
`when`	`(when c b ...)` equivalent to `(if c (do b ...))`	`(when is-sending (+ acc 1))`
`unless`	`(unless c b ...)` eqv. to `(if (! c) (do b ...))`	`(unless overflow "ok")`

`cond`	Tests multiple conditions and evaluates first sat. condition
	`else` can be used as a default condition
		`(cond [sending (+ acc data)]`
		`[receiving (print "Received:" data)]`
		`[else (print "waiting")])`

`while`	Evaluates body as long as condition evaluates to true	`(while (! receiving) (print "."))`

do	Evaluates all expressions in body. Result is last expr.	(do (set! x (+ 1 2))
		    (print x)
		    (+ (calc-delay) x))

defun	Defines a new function	(defun times-two [x] (* 2 x))
defmacro	Defines a new macro	(defmacro rev-args [xs] ‘(,(first xs) ,(reverse (rest xs))))
fn	Defines a new anonymous function	((fn [x] (* 2 x)) 4)
print	Prints the arguments to the standart output	(print "hello" (+ 1 2))
printf	Prints the arguments to the standart output depending on format string	(printf "%s: %d" "Res" (+ 1 2))
list	Evaluates arguments and put into new list	(list 1 2 (+ 1 2) "abc")
length	Returns the length of a list	(length (list 1 2 3))
min, max	Returns the min or max value of a list	(min (list 1 2 3 4))
sum, average	Returns the sum or average value of a list	(sum (list 1 2 3 4))
map	Creates a new list by applying a function to each element in a list. Function must take exactly one argument	(map (fn [x] (* x 2)) (list 1 2 3))
reverse	Returns the passed list in reversed order	(reverse '(1 2 3))
array	Creates a new array, entries can be passed as tuples	(array ["a" 1] ["b" (+ 1 1)])
seta	Updates value in an array	(seta array-var "b" 5)
geta	Gets value from an array	(geta array-var "b")
geta/default	Gets value from an array, if key not in array returns default value	(geta/default array-var "default" "b")
mapa	Returns a list by applying function to every key-value pair in an array. The function must take 2 arguments	(mapa (fn [k v] (* 2 v)) array-var)
resolve-scope	Takes the argument and appends it to the current scope to form a full signal name	(resolve-scope signal)
	The shorthand for resolve-scope is ~	~signal
in-scope	Evaluates the body expression in a scope. Inside signals can be scope-resolved	(in-scope 'top.sub ~ready)

(continued)

Table 5.2 (continued)

Name	Description	Example
`in-scopes`	Evaluates the body expression in all scopes	`(in-scopes '(top.sub1 top.sub2) ~ready)`
`resolve-group`	Takes the argument and appends it to the current group to form a full signal name	`(resolve-group signal)`
	The shorthand for `resolve-group` is #	`#signal`
`groups`	Returns a list of all groups that contain the signals passed as arguments	`(groups clk)`
`in-group`	Evaluates the body expression in a group. Inside signals can be group-resolved	`(in-group 'tb.dut (print #clk))`
`in-groups`	Evaluates the body expression in multiple groups	`(in-groups (groups clk) (print #clk))`
`find`	Returns a list of all indices at which expression evaluates to true	`(find (&& top.ready top.valid))`
`count`	Counts at how many indices the expression evaluates to true	`(count (&& top.ready top.valid))`
`whenever`	Evaluates the body expressions at each index where the condition is true	`(whenever (&& a (! b)) (print a))`
`slice`	Returns a slice of bits from a signal.	`(slice top.sig 1 0)`
	The shorthand for `slice` is `[h:l]`	`top.sig[1:0]`
`timeframe`	Evaluates body expressions in a new local timeframe.	`(timeframe (step 10)`
	When the timeframe is exited, INDEX is restored	`(print INDEX))`
`sample-at`	Sets the sampling points to the indices passed as an argument	`(sample-at (find (&& (! clk@-1) clk)))`

the waveform data from Fig. 5.1 has been dumped to the file `"waveform.vcd"`. As a first step, we load this file into WAL under the id `w` as following: `(load "waveform.vcd" 'w)`.

After a waveform has been loaded, its time index is set to the beginning to 0, and it is available to WAL expressions. The `step` operator can be used to step the time index forward and backward by a variable amount.

For example, `(step 2)` increases the time index of all loaded waveforms by 2, while `(step -1)` decreases the time index of all loaded waveforms by 1.

5.3.3.2 Signal Access

After loading the waveform containing the data in Fig. 5.1, we can start writing our WAL solution to determine the average latency. In a first basic version of this program, we want to detect when `comp1` requests the bus and when there is the corresponding acknowledgment. As mentioned before, waveform signals are first-class citizens of the WAL language. Therefore, to access the signal value at the current time index of a waveform, it is sufficient to write the full signal name (i.e., a global name of the form `top.sub.signal`).[2]

In our problem case, a request is said to be acknowledged when both the `req` and `ack` signals are high. This condition can be expressed by a Boolean conjunction of the signals `comp1.req` and `comp1.ack` using the following WAL expression: `(&& comp1.req comp1.ack)`.

In the same way, we describe pending requests using the next WAL expression, when the `req` signal is high but the `ack` signal is low:`(&& comp1.req (! comp1.ack))`. This time, the signal `comp1.ack` is inverted using the `!` function, since the component has not yet processed the request and thus `comp1.ack` is set to 0.

If multiple waveforms are loaded, signal name ambiguities must be resolved by specifying the waveform id in front of the full signal name (e.g., `w$comp1.req` vs. `w2$Top.sig`). The id in front of the name can be omitted if only one waveform is loaded.

5.3.3.3 Average Latency for Component 1 in WAL

Now, we can combine the presented WAL functions to solve the average latency problem of Sect. 5.2 for `comp1`. The WAL program is shown in Listing 5.1. First, on Line 1, the waveform is loaded. As we are interested in the average latency w.r.t., the complete waveform, on Line 4 we step forward[3] until the end of the waveform

---

[2] It is also possible to extract specific bits from a signal value using slicing functions.

[3] The step size is 2 since we only want to sample data at positive clock edges which in this trace are located on every other index.

```
1 (load "waveform.vcd")
2 (define packets 0)
3 (define wait 0)
4 (while (step 2)
5   (when (&& compl.req compl.ack) (inc packets))
6   (when (&& compl.req (! compl.ack)) (inc wait)))
7 (print (/ wait acks))
```

Listing 5.1 Average Latency for compl

```
(&& (&& comp1.req (! comp1.ack))
    (&& comp1.req (! comp1.ack))@1
    (&& comp1.req (! comp1.ack))@2)
```

Listing 5.2 Detecting continuous pending requests

is reached. The "core detection" of requests and acknowledgments is performed on Lines 5–6. To compute the average latency, we have to determine the number of all acknowledged packets. This is done on Line 5, where the variable `packets` is incremented when a request is acknowledged. For this condition, we inserted the previously introduced expression for acknowledged requests. On Line 6, the `wait` variable is incremented when the component has a pending, unacknowledged request using the other previously introduced expression. Finally, after the end of the waveform has been reached, we calculate the average latency using division and print it to the standard output on Line 7.

5.3.3.4 Timing

Often, interesting signal relations are not limited to a single time index. For example, detecting a value change on a signal requires observing two values of the same signal at different time indices. This could be achieved by temporarily storing the first signal value in a variable, but this quickly becomes inconvenient. WAL overcomes this problem by allowing to modify the time index of a waveform locally for a specific expression. For this, we introduce the *relative-eval* operator `reval` which takes a target expression and an offset expression that must evaluate to a signed integer and evaluates the target expression with a locally changed time index according to the evaluated offset expression. The integer specifies the time offset at which the expression is evaluated relative to the current time index. For instance, detecting a signal value change can be expressed using the following WAL expression: `(!= (reval sig -1) sig)`. As relative evaluation is commonly needed, it can be abbreviated by appending an `@` followed by an offset to any expression. Using this shorthand syntax, the expression `(reval sig -1)` can be written as `sig@-1`.

As an example, we assume HW designers have to check a worst-case requirement and therefore have to find pending requests that are persisting at least three consecutive cycles. This can be expressed as shown in Listing 5.2.

5.3.4 *Macro System*

One of the defining features of programming languages in the Lisp family is that most of them have very sophisticated macro systems. Compared to macros in other languages, they are more than mere text substitution, and since they work on S-expressions, they can operate on the real structure of a program. Additionally, at expansion time, Lisp macros can utilize the full language to generate code as they can be thought of as regular functions that just return program code which is inserted in place of the macro call. This powerful macro system can be used to expand the language to ones needs, for example, to define new language constructs. We will give examples for this in the sections about the WAL's new standard library and WAL timeframes, respectively.

Before a WAL expression is evaluated, it is analyzed if it contains any macro applications. If this is the case, the macro function is called, and the macro application is replaced by the result of the macro function.

Using the `defmacro` keyword, new macros can be defined. Defining macros is very similar to defining functions, as the `defmacro` function also expects a name followed by a list of arguments, while all additional arguments form the body of the macro.

Listing 5.3 shows a macro that can be used to step forward until a condition is true. Since in WAL code and data are represented by the same data structure,[4] generating code is as simple as generating some nested lists. It is important to think about when an expression should be evaluated. If it should be a part of the code that is generated by the macro and not be evaluated instantly, it must be *quoted* using the `quote` function or its shorthand `'`. For example, we have to quote the `(step)` expression on Line 4 or else the step function would be evaluated immediately leading to a wrong INDEX (see also Sect. 5.3.5.3). However, evaluating `(step)` like this during the macro expansion is not what we want, and therefore we quote it to put the expression as a datum into the list. This way, we are able to place the `(step)` expression in the code the macro generates which is now evaluated only when the generated code is evaluated and not at expansion time. This resulting list can now be evaluated just as if it would be read from a WAL file.

Sometimes, it is useful for debugging to expand a macro without evaluating the resulting code. This is possible using the `macroexpand` function. For example, expanding `(step-until overflow)` results in the expression `(while (&& (! overflow) (step)) INDEX)`.

Since WAL expressions are represented by lists, macros that expand to more complex expressions generate deeply nested list structures. The above macro definition is very short. However, the expression generated by the macro is hard to read, since all nested lists are constructed manually. Thus, the macro contains many calls to the `list` function and many required quotations. For example, if we want to

[4] This is called homoiconicity.

```
(defmacro step-until [condition]
  "Step forward until condition is true"
  (list 'while (list '&& (list '! condition)
                          '(step))
                 'INDEX))
```

Listing 5.3 Defining WAL macros with the defmacrofunction

```
(defmacro step-until [condition]
  `(while (&& (! ,condition) (step)) INDEX))
```

Listing 5.4 The step-until macro with quasiquote

generate an expression of the form (+ x (* y z)), where x, y, and z are variables that contain other expressions which should be inserted into the expression, we would have to create a nested list by hand using following expression: (list '+ x (list '* y z)). This is hard to read, since it contains noise (e.g., the list function calls) and it does not look like the desired result (e.g., the + operator is not the first element in the list).

To improve the readability of macro definitions, WAL implements the quasiquote function known from other Lisp languages. A quasiquoted expression is evaluated like a regular quoted expression except that sub-expressions can be evaluated using the unquote function. Similar to the regular quote function, quasiquote and unquote have shorthands (c.f. Table 5.1). Using those two functions, the previous nested list can be generated with ‘(+ ,x (* ,y ,z)). The *step-until* macro can now be defined much more readable, as shown in Listing 5.4. Here, the quasiquoted list will be returned as is, with the exception that the condition expression will be evaluated and replaced by the result that was computed.

5.3.4.1 Reducing WAL Core Using Macros

Using the macro system, we can now implement a large part of WALs functionality as WAL macros and functions. Instead of implementing these functions inside a WAL interpreter, they are now implemented in a standalone WAL program that is evaluated before a WAL program is run. This *WAL standard library* significantly reduces the complexity of WAL interpreters, and hence WAL core, since the number of functions that have to be implemented can be reduced.

Listing 5.5 shows two examples for WAL functions that were previously implemented inside the interpreter, but which are now implemented as macros in the WAL standard library file. The first macro, count, counts how often a condition is true on the complete waveform by combining the length and find functions of WAL. It is expanded to calculating the length of the list returned by a call to the find function with the given condition.

```
(defmacro count [condition]
  `(length (find ,condition)))

(defmacro cond args
  (fold
    (fn [acc branch]
      (let ([condition (if (= branch[0] 'else)
                               #t
                               branch[0])]
            [then (rest branch)])
          (if acc
              `(if ,condition (do ,@then) ,acc)
              `(if ,condition (do ,@then)))))
    '()
    (reverse args)))
```

Listing 5.5 Exemplary macros from the WAL standard library

```
(cond [(= state 0x8) (print "Sending")]
      [(= state 0x4) (print "Received:" data)]
      [else (print "waiting")])
```

Listing 5.6 Examplary usage of the cond macro

```
(if (= state 0x8)
    (print "Sending")
    (if (= state 0x4)
        (print "Received:" data)
        (print "waiting")))
```

Listing 5.7 Expansion of the condexpression in Listing 5.6

The second macro, cond, constructs nested *if-else* expressions by iterating over the arguments given to the cond macro. These arguments are lists where the first element is a condition and the rest are expressions that get evaluated if the condition evaluates to true. This macro is a good example to showcase how macro expansions can take advantage of the full WAL language as it uses several advanced WAL functions such as `fold`, lambda functions, and variable assignments to generate the resulting code.

An exemplary usage of the cond macro is shown in Listing 5.6. There, the cond macro is used to print a log message based on a *state* register. Listing 5.7 shows the nested if-else expressions generated by the cond macro. Every condition of the cond macro is translated into an if-else expression except the last one where the *else* condition is translated into an always true statement. Therefore, the if expression can be optimized away.

5.3.5 Advanced WAL Operations

The expressiveness of WAL, based on the essential operations as introduced in the previous section, is sufficient for a wide range of analysis problems. However, the applicability of WAL can be significantly improved by adding advanced features. This allows to write much more compact, more generic, and much easier to read WAL programs.

5.3.5.1 Calling External Code

WAL enables developers to write concise, powerful, and easy-to-use programs for waveform analysis. On the other side, many problems not related to waveform analysis (such as UI or databases) are already available in libraries for other programming languages. Combining WAL with other programming languages saves time and helps to integrate WAL into complex workflows. Therefore, WAL enables users to tap into the large ecosystems of other programming languages. Using the `import` function, external code in another language[5] can be imported into running WAL programs. After importing, external functions can be called using the `call` function.

5.3.5.2 Logical Grouping

Our example design contains two components that are connected to the bus, `comp1` and `comp2`. Both components are structurally similar w.r.t., the bus communication interface (e.g., same signals). Coming back to our example, the problem *"How is the latency of* `comp1`*?"* is also valid for `comp2` or any other component attached to the bus. An elegant solution requires the separation of the core problem and the concrete signal names. WAL supports writing these separated *generic* expressions through a set of concepts and functions.

First, we introduce the concept of a *group*. A group is a set of signals which are semantically connected (e.g., the signals of a bus). Groups are defined by a prefix (a partial signal name) and a set of postfixes for which the combination *prefix* + *postfix* results in an existing signal name for every postfix. For example, the waveform in Fig. 5.1 contains two groups, `"comp1."` and `"comp2."`, for the set `req` and `ack`. Users can search the design for groups using the `groups` function (i.e., `(groups "req" "ack")` for the example).

To make use of a group, it has to be *captured* first. Capturing a group defines this group as the current active group and allows accessing signals in the group using just the postfixes. Groups are captured using the `in-group` operator, which takes

[5] This is, for example, code in the host language of the WAL interpreter.

a `group` and then evaluates the `body` expression. The `in-groups` function works in the same way but expects a list of groups and evaluates the body expression once for each of these groups. During the evaluation of the body expression, the specified group is marked as the current group (the active Current Group is available using the `CG` special variable). Accessing signals in a group just by a postfix is called *resolution*. In the body of an `in-group[s]` expression, signal names can be resolved using the `resolve-group` function. This function takes a symbol and appends it to the captured group. If the resulting symbol refers to an existing signal, the value of this signal at the current time is returned. As wrapping all signals in `resolve-group` function calls leads to increased verbosity, a shorthand for this function is to add a # in front of a symbol (cf., Table 5.1).

WAL expressions can make use of the hierarchical information of waveform data. The *scoping* concept allows evaluating WAL expressions in selected scopes (i.e., the submodules of the design). Scoping is available via the $\sim$ shorthand and the scoping functions. Since scoping works similar to grouping, we omit further details.

5.3.5.3 Timeframes

In WAL, each trace has a global `INDEX` that points into the trace and thus influences which value is returned when a signal is evaluated. Often however, programs need to modify the `INDEX` while walking over the trace but then need to restore the original `INDEX`. Of course, this behavior can also occur nested arbitrarily deep, which in turn results in the creation of a stack-like structure. This functionality could be implemented without additional functions by manually storing and restoring the current `INDEX`. However, WAL aims to make describing analysis problems as easy and the analysis code as descriptive as possible.

Take, for example, a bus on which multiple transactions can be "live" at the same time, i.e., a new request can be generated before the previous transactions are acknowledged. For this, each transaction is tagged with a transaction id.

A WAL program to analyze a system like that is shown in Listing 5.8; please ignore the `timeframe` function on Line 2 for now. This program visits every time point at which a new transaction is seen on the bus. It then stores the transaction id and steps forward until the transaction with this id is acknowledged and prints a debug message. Now, after the program reaches the time at which the transaction is acknowledged, it returns to the outer `whenever` loop. Yet, as the `INDEX` was modified inside the previous iteration, it does not continue at the time point after the `bus_ready` signal was high but after the transaction was acknowledged. This is a problem, since new transactions could have been started between these two time points.

To handle situations like this in a convenient way, we propose so-called timeframes. A `timeframe` is an environment that stores the current `INDEX` for each loaded trace (accessible inside the timeframe via `TIMEFRAME-START`), evaluates its body expressions, and then restores the indices back to their original values. Now, the

```
(whenever bus_ready
  (timeframe
    (let ([id bus_id])
      ;; step until next ready
      (print "Request from " INDEX)
      (step-until (&& bus_ack (= id bus_id)))
      (print " to " INDEX))))
```

Listing 5.8 By using a timeframe, step-until does not interfere with the whenever function

problem in Listing 5.8 can be solved by wrapping the whenever body inside a timeframe. By this, the INDEX is restored back to its original value of when the body of the whenever function was entered.

5.3.5.4 Modifying the Sampling Points

WAL makes no assumptions about the semantics of signals. Therefore, for example, clk or reset signals have no special meaning that change the behavior of WAL. However, a lot of analysis problems are best formulated using the clock-sampled synchronous semantics known from RTL modeling or from assertion languages. Please recall that the INDEX is incremented, whenever a value changes inside a trace. Compare this to clock-sampled semantics where users might expect that the INDEX is only incremented, for example, on each new rising clock edge. This also influences the semantics of relative evaluation. For example, signal@-1 should now return the value which signal had at the last clock cycle.

To achieve this, we present the sample-at function which changes at which time points signal values are sampled from the trace and how the INDEX is calculated. This function expects a list of indices that has to be a subset of all indices of the loaded trace. After sample-at is called, signals can only be read at time points that are included in this subset of indices. Additionally, the new time points are given new indices starting at 0. This process of changing the sample points of a trace is shown in Fig. 5.3. The trace with the original sample points above the waveform is shown in Fig. 5.3a (see red arrows). Please note how each change of a signal introduces a new sample point and a new index (see INDEX 2 for an asynchronous new index). In this example, we want to sample from the trace only when the clk signal is high. Therefore, we get a list of all indices at which this condition is true by evaluating (find (= clk 1)). The indices that are contained in the returned list are shown in Fig. 5.3b. Finally, by passing this list to the sample-at function, we update the sample points of the trace which results in the new indices shown in Fig. 5.3c. Now, the expression data@-1 has the meaning of the value of the data signal when the clock was high the previous time instead of the value of the data signal at the time of the last signal change.

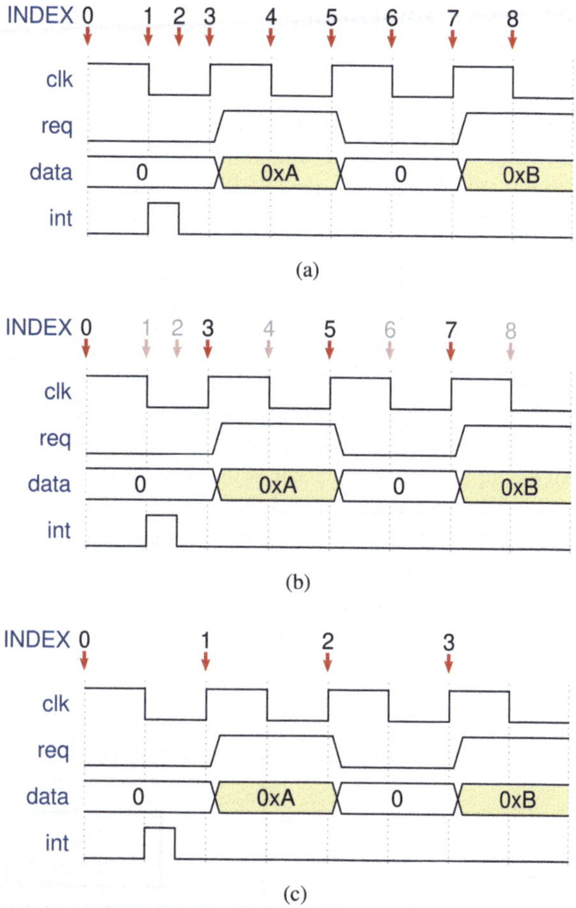

Fig. 5.3 The sampling points of a trace can be changed with the `sample-at` function. In this case we sample only on rising clock edges using the expression `(sample-at (find (= clk 1)))`. (**a**) Original sample points. (**b**) New sample points. (**c**) Updated samplepoints and new indices

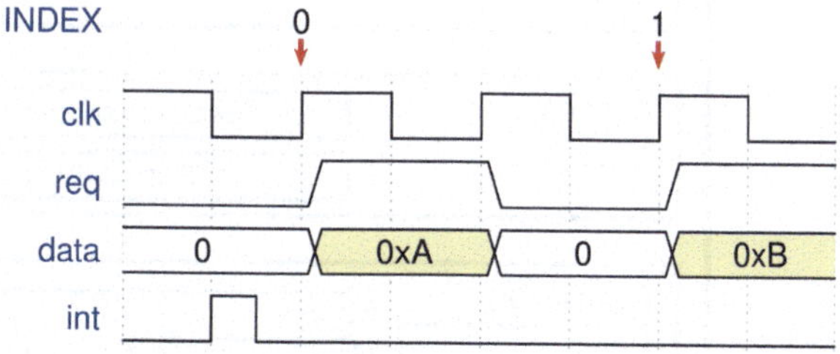

Fig. 5.4 Sampling at `(&& (= clk 1) (= req 1))`

The `sample-at` function is not only limited to sampling at clock edges. For example, in Fig. 5.4, we sample only when both `clk` and `req` signals are high, thus, modeling a valid transaction on a simple bus. Now, the expression `data@-1` has the meaning of the `data` signal at the last valid transaction.

5.4 WAL Interpreter

We have implemented a Python-based interpreter for WAL which we refer to as the *WAL core* . WAL core is a Python package that contains an API for WAL integration in Python applications, a standalone interpreter to run WAL programs in a terminal, and a Read-Eval-Print-Loop (REPL) shell for interactive WAL programming. WAL core can analyze waveforms in the VCD and *FST* [3] formats.

The architecture of WAL core is depicted in Fig. 5.5. The inputs to WAL core are user-defined analysis programs (see top of the figure). These programs are either WAL programs or programs in a different language that are transpiled to WAL. In the first case, the WAL core *Reader* transforms WAL programs from text form to *Internal Code*. The same flow via the Reader is also possible for a *Custom Front-End*. Alternatively, a front end has the option to utilize its own reader, generating WAL internal code directly without relying on the Reader module in the WAL core. An example of this approach is our Waveform AWK (WAWK) language which is presented in detail in Sect. 5.5.5.

Next, the internal code is run through passes. In Fig. 5.5, these are *Macro Expansion* followed by *Optimization*. The macro system embedded within WAL empowers users to construct complex operations by composing simpler operations, yielding numerous advantages across all language levels. Furthermore, this feature facilitates the creation of libraries, amplifying the language's capabilities.

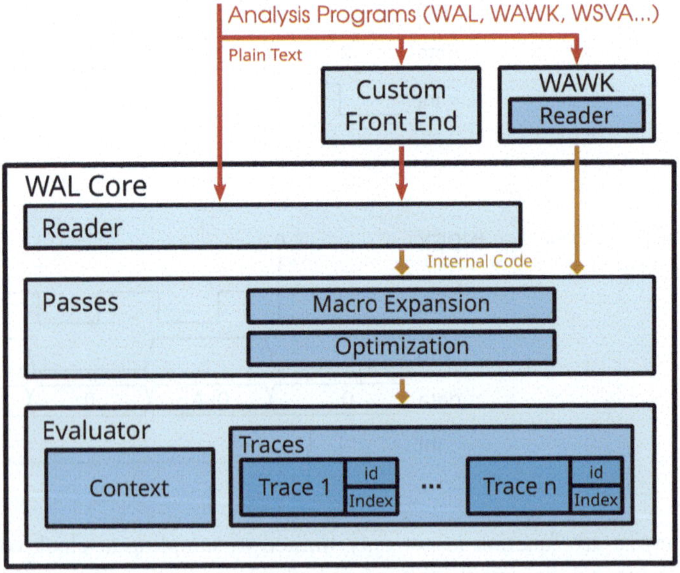

Fig. 5.5 Architecture of the WAL interpreter (referred to as WAL core)

The result of the passes is then evaluated, i.e., executed, by the *Evaluator*. The evaluator is also the place where the execution state is stored. The execution state consists of the loaded traces and the context which contains all the defined variables.

WAL core is not limited to just executing WAL program files; it also allows embedding WAL into Python programs. Using the API of WAL core, non-trivial analysis tasks can be easily performed in Python applications, which we demonstrate in Sect. 5.5. In addition, WAL core provides access to the internal WAL structures, which in combination with the API, enable a clean interface for WAL-IR applications. Currently, WAL core is not optimized for performance. Instead, it is intended as an executable specification that is both easy to understand and easy to extend to enable further research.

Further, with the macro system we presented in Sect. 5.3.4, we shift more functionality out of WAL core toward the standard library. This makes WAL core even smaller and provides a lot of functionality defined in WAL itself which is now ready to use in alternative WAL implementations.

5.5 Case Studies

In this section, we explore case studies that represent different WAL use cases. First, in Sect. 5.5.1, we expand the scope of the analysis to the profiling of software running on RISC-V processors using WAL and the symbol information contained in the application binaries. Next, in Sect. 5.5.2, we present an *APB* monitoring function which can be used to perform various tasks by providing different callback functions. Then, in Sect. 5.5.3, we introduce a DSL for specifying processor pipelines which we use for a web-based pipeline viewer. Finally, in Sect. 5.5.4, we utilize WALs macro system to create a new SQL inspired DSL which is embedded in WAL and allows gathering large amounts of data with simple queries.

5.5.1 Waveform-Based Profiling of RISC-V Binaries

In this section, we present how RISC-V binaries, that are executed on simulated cores, can be profiled using WAL. For this, we utilize the symbol information contained in the binary and match this information against the instructions which were executed during simulation. The profiling is implemented as a *Python application* (an excerpt of this application is shown in Listing 5.9) that handles both the extraction of symbol information and the waveform analysis using WAL. First, the application extracts the address ranges of functions from a given ELF binary using the *nm* command from the RISC-V toolchain (this happens in the function *ranges* on Line 1 of Listing 5.9 which we omit for brevity). All information about the functions is stored as triples consisting of the function name, the start address, and the end address of the function.

Next, the WAL interpreter is instantiated on Line 3, and the simulation trace is loaded on Line 4. The analysis is processor independent and relies on two definitions (which can be implemented as either functions or macros) that must be implemented for a specific core, *fire* and *pc*. The *fire* definition, when evaluated, should return true when the core finishes an instruction at the current `INDEX` and false otherwise. The *pc* definition, when evaluated, should return the program counter value that is associated with the currently finished instruction. Listing 5.10 shows macros that implement the required functionality for the *VexRiscv* processor. The core of the profiling analysis is the function *count-function* (Lines 8–13 in Listing 5.9) which matches the current program counter against the function address ranges read from the binary. If the program counter lies within a functions address range, the counter associated with this function in the `dist` (short for function distribution) array is incremented.

This `dist` array is instantiated in the Python code on Line 17 and passed as an argument to the WAL interpreter when the analysis is performed (Line 22). By passing the array into the interpreter like this, it is placed in the execution environment and the WAL code can directly work on it. This showcases how WAL and Python code can interact with each other, thus enabling developers to partition the code such that each subproblem can be implemented in the most natural language.

The main analysis loop is shown on Lines 18–22 of Listing 5.9. This expression traverses the full trace calling `count-function` whenever an instruction is finished by the processor. Additionally, it also counts the number of instructions that have been executed overall. This value is returned to the Python code, since it is the result of the `whenever` expression, which is the last evaluated statement of the body.

Listing 5.11 shows an excerpt of the profiling results, listing the function, its address range, and the number of times an instruction from this range was executed.

5.5.2 Analyzing APB Interfaces

The Advanced Peripheral Bus (APB) protocol is a widely used communication interface for low-speed devices. It is a synchronous protocol with two independent buses, one for reading and one for writing, which, however, cannot be used at the same time.

In this section, we present a monitor function which can detect events (e.g., the end of a successful write transaction) on an APB interface. Users can specify a callback function for any of the supported events, which is called if this event is detected. By plugging in different kinds of callback functions, the general APB monitor can be used to perform a wide range of tasks.

The code for the APB monitor function is shown in Listing 5.12. Five parameters are expected by the `monitor-apb` function, the group of the APB bus which should be analyzed and four callback functions.

```
functions = ranges(BIN)

wal = Wal()
wal.load(VCD)
# config script for core-specific names
wal.eval_str('(eval-file config)')
wal.eval_str('''
(defun count-function [addr]
  (for [f funcs]
    (when (&& (>= addr f[1]) (<= addr f[2]))
      (seta dist
         f[0]
         (+ (geta/default dist 0 f[0]) 1)))))
''')

# calculate the time spent in each function
dist = {}
instruction_executed = wal.eval_str('''
(whenever (fire)
          (count-function (pc))
          (inc ninstr))''',
  funcs=functions, dist=dist)
```

Listing 5.9 Python code for function profiling using WAL

```
(defmacro fire []
  `(&& (rising TOP.clk)
       (! TOP.reset)
       TOP.VexRiscv.lastStageIsFiring))
(defmacro pc [] 'TOP.VexRiscv.lastStagePc)
```

Listing 5.10 VexRiscv specific code for the function profiling

```
bss_loop    [0x800000cc-0x800000d8]: 11273
puts        [0x80000e38-0x80000e78]: 9527
printf      [0x80000d34-0x80000e34]: 7362
__divsi3    [0x80002ec4-0x80002f44]: 6344
printf_s    [0x80000c40-0x80000c74]: 1565
printf_d    [0x80000c7c-0x80000d08]: 1325
main2       [0x800002f8-0x80000a78]: 400
__mulsi3    [0x80002ea0-0x80002ec0]: 343
...
```

Listing 5.11 Excerpt of profiling results

```
1  (defun monitor-apb [g setup wait read write]
2    (in-group
3      g
4      (whenever (&& (rising #pclk) (= #presetn 1))
5        (case (list #psel #penable #pready)
6          [(1 0 0) (when setup (setup))]
7          [(1 1 0) (when wait (wait))]
8          [(1 1 1) (if #pwrite
9                     (when write
10                      (write #paddr #pwdata))
11                    (when read
12                      (read #paddr #prdata)))]))))
```

Listing 5.12 Generic APB monitoring function

Lines 4–12 contain the main logic of the function. Since APB is a clock synchronous protocol, the function visits every timestamp at which the APB clock is rising and the active-low APB *resetn* signal is high. At each timestamp at which this condition is true, the state of the bus can be detected by observing the `psel`, `penable`, and `pready` signals. On Line 5, the current values of these signals are sampled and collected into a list. Then, we pattern match this list using the `case` function against three constant lists that represent the *setup*, *wait*, and *enable* states of an APB bus. If the sampled signals now match one of the three lists, the corresponding callback function is executed, but only if a valid callback function for this event is available. Only for the *enable* state on Lines 8–12, the code further checks if the transaction is a read or a write access. The main benefits of this pattern matching approach are that it reduces the number of nested if expressions and that it plainly documents the possible types of APB events. However, since the monitor function is not a protocol checker, the pattern matching also requires that the bus adheres to the specification.

Listing 5.13 presents three exemplary use cases for the `monitor-apb` function which we explain in the remainder of this section.

5.5.2.1 APB Stream Logging

First, in Listing 5.13 on Lines 1–6, a macro is defined which can be used to print a report of the transaction on the specified APB bus `g`. The macro expands to an application of the `monitor-apb` function, in which only the `read` and `write` callbacks are used. For them, two lambda functions are supplied that print a string with information about the transaction. The other callbacks are disabled by supplying them with zeros instead of functions.

5.5.2.2 APB Slave Memory Restoration

Often, memories are not dumped into waveforms due to space and efficiency reasons. Therefore, the memory module acts as a black box, since the waveform never has its full state. However, by monitoring the transaction on a bus, it is possible to reconstruct the memory from a certain initial state. Lines 8–12 in Listing 5.13 shows a function that, if passed the group to an APB bus, restores the memory content based on observed write transactions. This is done via the array defined on Line 9. This array keeps track of the current value of every address and is updated in the `write` callback on Line 11.

5.5.2.3 APB Transaction Delay

Lines 14–25 in Listing 5.13 present a function which, similar to the earlier example, calculates the average delay on a given APB bus. This is implemented using the `setup`, `write`, and `read` callbacks. The logic behind this code is that the current timestamp (TS) is stored in the `start` variable whenever the APB bus enters the *setup* phase (see `on-setup` function on Line 17). When either a read transaction or a write transaction is detected, the time that passed since the last setup phase is calculated (Lines 18–19) and appended to a list. Finally, the average of the list is computed and returned as the result of the `apb-avg-delay` function. Please note that by providing the `on-enable` function to only the `read` or `write` callbacks, the delay analysis can be performed for only one transaction type if required.

5.5.3 A DSL for Specifying Pipelines

For a few years now, RISC-V development boards in the embedded category are available, new open-source cores are published regularly, and work is underway to spread RISC-V into more and more domains [6, 35]. Recently, even the first low-cost RISC-V development boards capable of desktop environments have arrived [34] bringing with them the potential to popularize RISC-V with even more developers. Yet, silicon alone is not enough to establish an ISA in the long run, but a large and active software ecosystem and first-class development tools are required as well.

On one hand, these advancements into wildly different domains are a manifestation of RISC-V's modularity and customization potential. On the other hand, it also means that, for RISC-V even more than for other ISAs, no two cores are alike, and thus it is increasingly complicated to provide excellent tool support for as many cores as possible. It also means that the tools must be as flexible as RISC-V itself.

In this case study, we present a DSL for specifying processor pipelines that can be used to extract information about the pipeline from simulation waveforms. We present a web application that can visualize RISC-V pipelines with only a simulation waveform and a few lines of declarations. Since our DSL is based on WAL, complex

```
(defmacro apb-print [g]
  `(monitor-apb ,g ∅ ∅
     (fn [addr data]
       (printf "%4d: R(%s)=%s\n" TS addr data))
     (fn [addr data]
       (printf "%4d: W(%s)=%s\n" TS addr data))))

(defun apb-restore [g]
  (define mem (array))
  (monitor-apb g ∅ ∅ ∅
    (fn [addr data] (seta mem addr data)))
  mem)

(defun apb-avg-delay [g]
  (define start TS)
  (define delays '())
  (defun on-setup [] (set! start TS))
  (defun on-enable [addr data]
    (set! delays (append delays (- TS start))))
  (monitor-apb g
               on-setup
               ∅
               on-enable
               on-enable)
  (average delays))
```

Listing 5.13 APB callback functions

pipelines can be easily declared, and the analysis can be enriched by fully fledged programs for further performing complex analyses on the waveform.

5.5.3.1 Defining Pipelines

In this section, we present an exemplary pipeline definition for a slightly enhanced version of the pipelined processor introduced in [27]. This processor has five pipeline stages consisting of *fetch*, *decode*, *execute*, *memory*, and *writeback* stages, and it implements forwarding using a hazard unit. Figure 5.6 shows a block diagram of the processor.

In our DSL, new stages are defined using the `stage` keyword. The behavior of each stage can be configured using a set of defined arguments. These arguments control, for example, when a new instruction enters a stage, when the stage is flushed, or they can compute additional data that should be logged for the stage. The specific possible arguments are:

value — The next value of the pipeline. If unspecified, the value of the previous stage gets sampled. Required for the first stage.

update — Controls when the stage gets updated. If unspecified, defaults to true.

```
(eval-file pipeline)

(stage fetch
       (value tb.dut.dp.instrf@1)
       (stall tb.dut.dp.stallf)
       (log stallf tb.dut.dp.stallf)
       (log pc tb.dut.dp.pcf))

(stage decode
       (update (! tb.dut.dp.stalld))
       (stall tb.dut.dp.stalld)
       (flush tb.dut.dp.flushd)
       (log pc fetch-pc@-1)
       (log rd tb.dut.dp.rdd)
       (log rs1 tb.dut.dp.rs1d)
       (log rs2 tb.dut.dp.rs2d))

(stage execute
       (update (! tb.dut.dp.flushe))
       (flush tb.dut.dp.flushe)
       (log pc decode-pc@-1))

(stage memory)

(stage writeback)
```

Listing 5.14 Pipeline Definition for processor from Fig. 5.6

- stall Controls when the stage is stalled. If unspecified, defaults to false.
- flush Controls when the stage is flushed. If unspecified, defaults to false.
- log Additional data that gets logged for this stage. Two expressions are required. The first expression is the name under which the data is logged, and the second is the WAL expression that is logged. The expressions can be arbitrarily complex WAL programs.

Listing 5.14 shows the pipeline definition for Fig. 5.6. Take, for example, the *decode* stage. This stage loads the next instruction when it is not stalled and when it is not flushed. In this case, the instruction from the *fetch* stage is loaded into the *decode* stage.

The definition of a pipeline is not linked to rendering it. By uncoupling data collection and presentation, both parts are now ready to be reused. Different rendering "back ends" can be selected, for example, text output or the web application we present in the following section.

5.5.3.2 Visualizing Pipelines on a Webpage

In this section, we present a web application that shows interactive visualizations of processor pipelines based on pipeline specifications in the previously presented

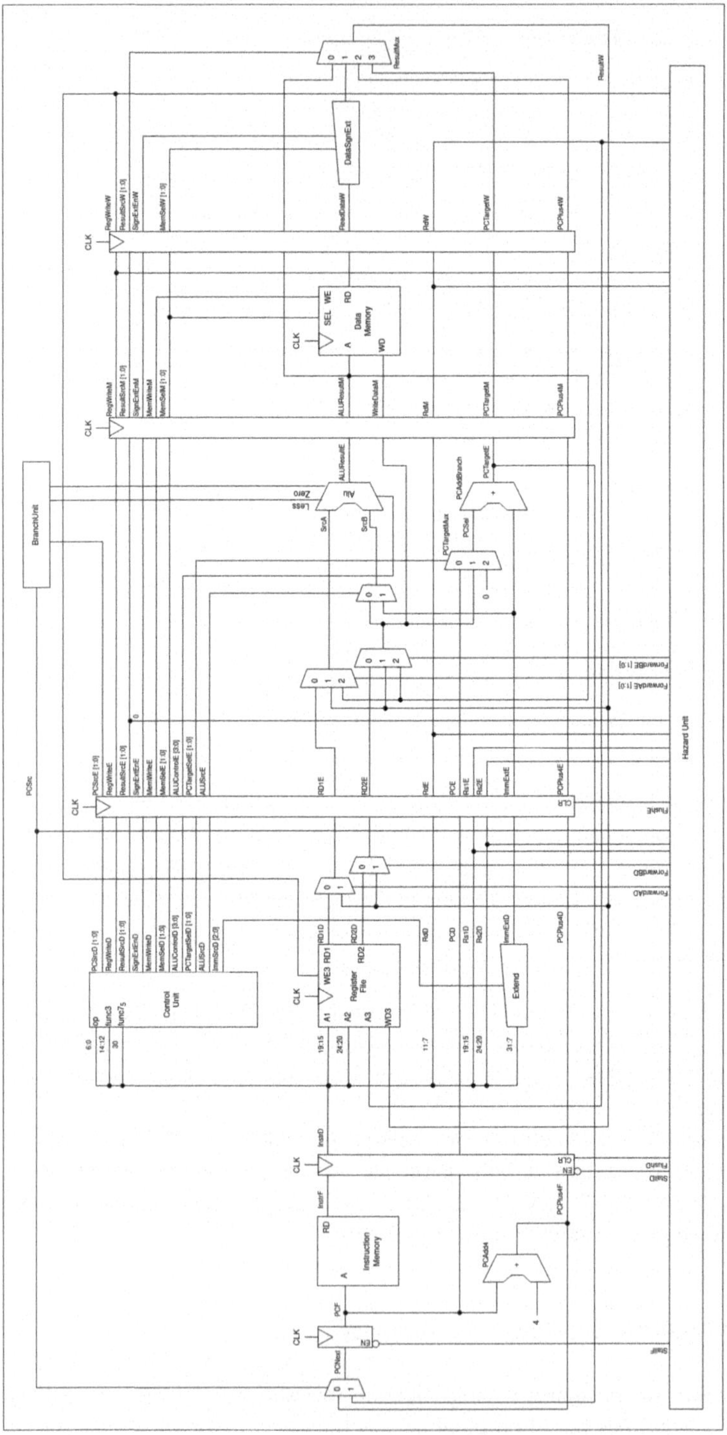

Fig. 5.6 Enhanced pipelined architecture from [27]

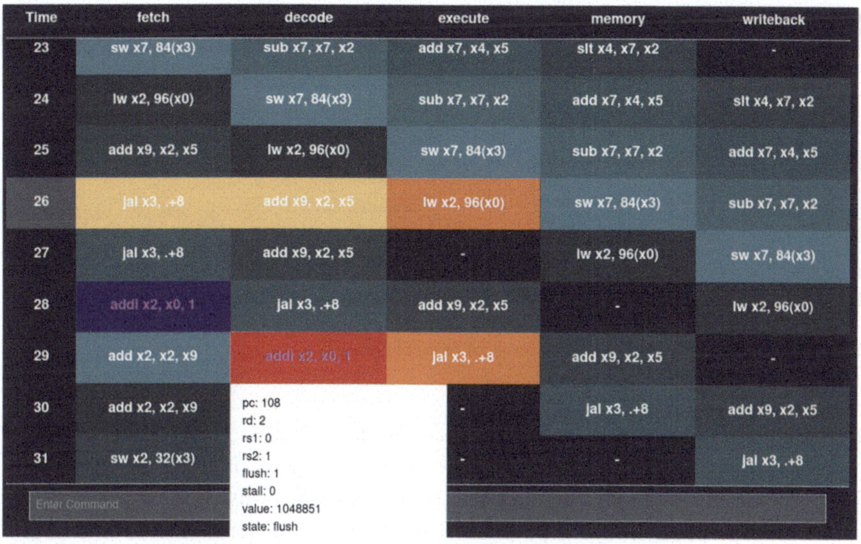

Fig. 5.7 Pipeline rendered as a website

DSL. The application is processor-independent, and the visualization depends solely on the pipeline specification made in our DSL.

Modern web technology allows highly dynamic and interactive applications, which is perfect for visualizing complex data such as processor pipelines. Additionally, it makes sharing and cooperating much simpler and allows working on mobile devices as well. After the pipeline is specified, a waveform can be processed to collect the information about all stages during the complete simulation (or parts of it).

Figure 5.7 shows our web application rendering the pipeline from Fig. 5.6 for a given waveform. Each column represents a pipeline stage, and the rows represent the time advancing with each clock cycle. Instructions are shaded differently to aid the eye while following an instruction through the pipeline. Stalls and flushes are marked in yellow and orange colors, respectively, while single instructions can be specifically highlighted in purple by hovering over it. The visualization can be navigated using the arrow keys and by entering special commands into an input field. Additionally, the pipeline can be searched to highlight cells depending on their contents (e.g., all stages with *add* instructions or all stages that operate on register x5).

5.5.3.3 Visualizing Pipelines in a Waveform Viewer

In addition to displaying pipelines inside a web browser, WAL also enables them to be shown in waveform viewers [18]. This is achieved by extending WAL with

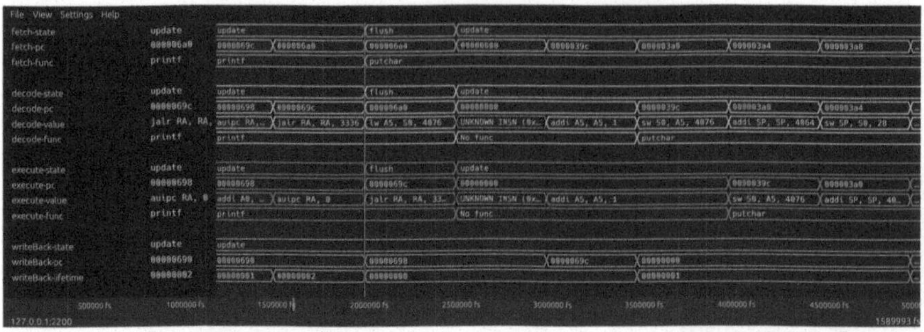

Fig. 5.8 Pipeline rendered in the Surfer waveform viewer

a protocol for waveform data transmission. Surfer is a waveform viewer [30] that has support for this protocol and can thus connect to a WAL interpreter. This allows Surfer to use WAL as the source for the displayed waveforms.

An example of displaying pipeline abstractions in Surfer is shown in Fig. 5.8. In this example, we use the pipeline DSL to create an abstraction of the processor pipeline shown in Fig. 5.6. The abstraction adds new signals which carry high-level information about the pipeline stages to the design (we will present more on how this is done in Chap. 8). For example, in Fig. 5.8, a signal named *stage-func* (with *stage* replaced by the stage name) is created for every pipeline stage using the pipeline DSL. These signals carry information about the currently executed functions, which is computed based on the current program counter value of the executing instruction and information extracted from the application binary. Further, *stage-state* signals indicate the current state of the pipeline stages. For example, when a stage is flushed, the *stage-state* signal will contain the value *"flush"*.

5.5.4 SELECT: An Exemplary DSL Built Using WAL Macros

The ease of defining new embedded DSLs using macros is one of the key benefits of using Lisp-like languages. They allow lifting a problem to a much higher abstraction level which can help solving it in a simple and efficient way. Further, DSLs created using macros are not to be used only in a standalone fashion but can be naturally embedded in host language programs.

In this section, we present a case study for SELECT, a DSL written as a WAL macro. This DSL presents a new, much more concise, and declarative syntax for reading the values of multiple signals at specified timestamps. In particular, SELECT is highly inspired by the syntax and clauses of SQL languages. The main idea behind this DSL is that declarative queries such as `SELECT s1 s2 FROM tb.dut.core1` can be used to get signal values from a waveform, without having to write WAL code that performs the required operations.

Select queries support the clauses `WHERE`, `ON`, and `LIMIT`. These clauses are parsed by the *SELECT* macro and get expanded into the required WAL code, thus hiding the complexity of writing the underlying WAL program to the user.

The code of the *SELECT* macro is shown in Listing 5.15. First, after the select clause, the macro reads the signals which should be contained in the result until another clause is detected (Lines 4–6). All clauses except the `SELECT` clause are optional and are initialized with sensible default values (e.g., when the `LIMIT` clause is omitted the number of rows returned is not limited) on Lines 8–11. The `resolve-signals` function is a helper function that takes a WAL expression and rewrites it such that all selected signals are resolved in the specified group. Then, `signaltuples`, a list of tuples containing the name and the value of all selected signals, is constructed. Afterward, the rest of the query is parsed, and the remaining default clauses are overwritten if they are specified (Lines 26–34). Finally, the code that implements the selection query is assembled (Lines 36–44) using the clause values parsed before.

Listing 5.16 demonstrates how the *SELECT* macro can be used to quickly get the values of some signals. In particular, the macro allows collecting quite complex data with very little code. In addition, the *SELECT* expression is much less deeply nested than the code produced by the macro and therefore much easier to write, especially in an interactive context like the WAL shell. `SHOW` is a utility function defined in the *SELECT* library which prints the result of a query as an ASCII table.
In general, the *SELECT* macro is only one example for a WAL DSL. Other examples could include a DSL that encodes processor instructions or one that encodes network packets. Depending on and utilizing the development context, custom DSLs can simplify expressing problems significantly.

5.5.5 WAWK: An AWK-Inspired Custom WAL Fronted

In this case study, we present a new DSL built on top of WAL using a custom front end (cf., Fig. 5.5). For this purpose, we introduce WAWK a new language inspired by the popular text-processing language *AWK*. We leverage the well-established programming paradigm of AWK to enable easy and concise programs for analyzing waveforms. Further, because WAWK uses a more common C-inspired syntax instead of WAL's S-expressions, it provides a lower entry barrier for people not familiar with programming languages from the LISP family.

In general, all WAWK programs consist of multiple statements that follow a `condition: { action }` scheme. At each time index in a waveform, the `condition` of every statement is evaluated, and, if satisfied, the associated *action* is executed. Additionally, two special conditions exist that allow executing actions before and after the waveform is processed. These are called `BEGIN` and `END`, respectively.

From the technical viewpoint, WAWK is a custom front end to WAL. This means that the interpreter parses a WAWK program using a *custom reader* (cf., Fig. 5.5).

```
(defmacro SELECT args
  (define signals '())
  (while (&& args
             (! (in args[0] '(FROM WHERE ON LIMIT))))
    (set! signals (append signals args[0]))
    (set! args (rest args)))

  (define group "")
  (define condition true)
  (define sampling true)
  (define limit true)

  (defun resolve-signals [e]
    (cond [(&& (list? e) (> (length e) 1))
           `(,e[0] ,@(map resolve-signals (rest e)))]
          [(&& (symbol? e)
               (in e signals)
            (! (in e '(TS INDEX SIGNALS CS CG))))
           `(resolve-group ,e)]
          [else e]))

  (define signaltuples
          (map (fn [x] `(',x ,(resolve-signals x)))
          signals))

  (while args
    (case args[0]
     [FROM  (set! group args[1])]
     [WHERE (set! condition (resolve-signals args[1]))]
     [ON    (set! sampling (resolve-signals args[1]))]
     [LIMIT (set! limit
                  (resolve-signals
                    `(< (length result) ,args[1])))])
    (set! args (rest (rest args))))

  `(let ([result '()]
         [default-sampling (find true)])
     (in-group ',group
       (sample-at (find ,sampling))
       (whenever (&& ,condition ,limit)
         (set! result
               (append result (array ,@signaltuples))))
       (sample-at default-sampling))
       result))
```

Listing 5.15 Macro for the query language

```
>-> (SHOW (SELECT INDEX counter
              FROM tb.dut.
              WHERE (> counter 3)
              ON #clk
              LIMIT 5))
+-----+-------+---------+
| Row | INDEX | counter |
+-----+-------+---------+
|   0 |    17 |       4 |
|   1 |    18 |       4 |
|   2 |    19 |       5 |
|   3 |    20 |       5 |
|   4 |    21 |       6 |
+-----+-------+---------+
```

Listing 5.16 Using the *SELECT* macro inside the WAL shell (>->is the WAL shell prompt)

It then directly maps WAWK constructs to *WAL internal code* and evaluates it using WAL core. By leveraging WAL core, we can implement the WAWK compiler in Python in less than 300 lines of code. Since the syntax of WAL is based on S-expressions, this mapping is made amazingly simple by combining existing WAL functions using the list utilities of Python.

In the following, we consider two exemplary use cases. First, we extract the basic blocks of software running on the VexRiscv processor [33]. Then, we analyze the runtime of instructions of the SERV RISC-V processor.

5.5.5.1 RISC-V Basic Block Extraction

For code analysis, we had to find the basic blocks of software running on the VexRiscv processor. A basic block is a sequence of continuous instructions that always execute in the given order, with no jumps in or out. We collected this information using a WAWK program that traces instructions and detects the beginnings of new blocks and the transitions between those blocks. The program in Listing 5.17 is composed of four statements.

Initially, *Statement 1* is executed once before the waveform is processed, which is specified by the special `BEGIN` condition. To improve the readability of the code, we define some shorter names for the most used signals using the `alias` function.

In *Statement 2*, the *pc* of the first block is stored in the variable *bstart* when *clk* and *fire* are set to 1 and *bstart* is 0 (which is true at the beginning since all variables are initialized to 0 at first use).

Statement 3 handles detecting the start of a new basic block. A new basic block starts if one of the following two conditions is met: (1) the last instruction was a jump instruction or branch instruction. Since the RISC-V ISA [36] encodes jumps and branches by setting the bits 6 and 5 of an instruction to 1, we test this condition on Line 13 and store the result in the variable `was_jump`. (2) the sequential execution

```
1  // Statement 1: create aliases
2  BEGIN: {
3    alias(fire, "TOP.VexRiscv.lastStageIsFiring");
4    alias(pc, "TOP.VexRiscv.lastStagePc");
5    alias(inst, "TOP.VexRiscv.lastStageInstruction");
6  }
7
8  // Statement 2: initialize pc
9  TOP.clk, fire, !bstart: { bstart = pc; }
10
11 // Statement 3: detect block boundaries
12 TOP.clk, fire: {
13   was_jump = (lasti[6:5] == 3);
14   run_into = (starts[pc] && (starts[pc] != starts[last]));
15   if (was_jump || run_into) {
16     trans[last, pc] = [last, pc];
17     ends[last] = last;
18     bstart = pc;
19   }
20   // save start of block
21   starts[pc] = bstart;
22   // save current instruction
23   last = pc;
24   lasti = inst;
25 }
26 // Statement 4: render graphviz file
27 END: { ...
28 }
```

Listing 5.17 WAWK program to extract and visualize software basic blocks

reaches an instruction that is already contained in a block. This is the case when `starts[pc]` is not 0 and the start of the current block is not the same as the block of the previous instruction (i.e., the instruction at `pc - 4` assuming an RV32 ISA). We test this condition and store it in the variable `run_into` on Line 14. On Line 15 we check if one of the above conditions is true, and if satisfied store a new block and block transition. A new transition from the previous block into the current block is stored in the associative array `trans` on Line 16. On Lines 17–18 we store the previous instruction as the end of a block in the associative array `ends` and set the current `pc` as the start of the current block. Next, on Lines 21–24, we mark the current instruction as belonging to the current block and save the current `pc` and `inst` for the next instruction.

Finally, *Statement 4* generates and prints the graphviz code to render the basic blocks and the control flow of the software. The condition of *Statement 3* consists of the special symbol `END` and thus is only executed once after the waveform was completely processed. For brevity, we omitted the body of *Statement 3*, since it only contains some loops and print statements to render the graphviz file.

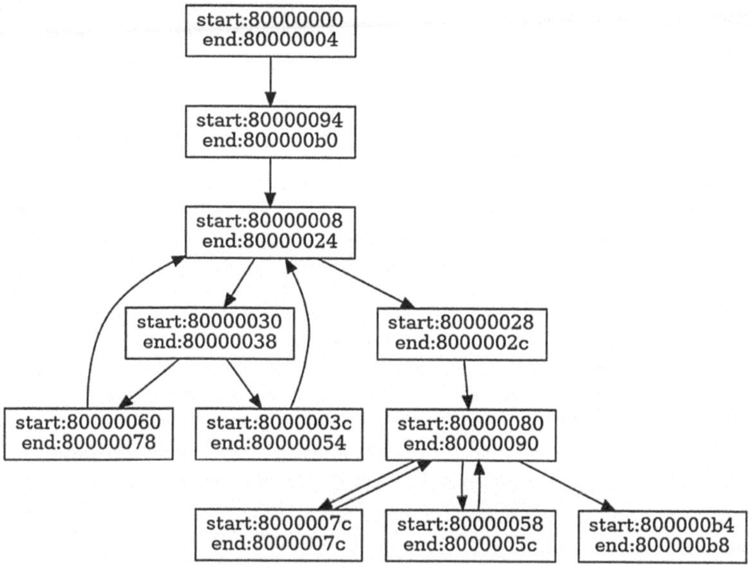

Fig. 5.9 Basic blocks for GCD calculation implemented in C

As an example, we analyzed the basic blocks of a basic C implementation of the greatest-common-divisor algorithm. The resulting basic block graph is depicted in Fig. 5.9. The WAWK program needed less than a second for processing the 4 MB large waveform file. To check how the program runtime scales with larger waveform files, we evaluated waveforms of the Dhrystone benchmark. The analysis of a medium-sized 265 MB waveform took around 42 seconds and found 207 basic blocks connected by 298 edges. For the second run, we increased the amount of Dhrystone iterations to produce a much larger waveform file, i.e., more than 4 GB. Running the same WAWK program as in the previous case took 586 seconds, and thus the run time scaled approximately linear to the waveform size.

5.5.5.2 Instruction Runtime Analysis on the SERV Processor

SERV [29] is a bit-serial implementation of the RISC-V with a heavy focus on very low area usage. The bit-serial implementation dictates that all instructions are executed bit by bit such that, for example, the RV32I `add` instruction is split over at least 32 cycles, one for each bit of the result.

We can observe the values of the instruction fetch bus to analyze the runtime of individual instructions in the SERV core. The start of a new instruction is signalized by the core when the signal `i_ibus_ack` rises. Therefore, the runtime of one instruction can be calculated by observing two following rising edge events on `i_ibus_ack` and calculating the time difference $time(event_2) - time(event_1)$

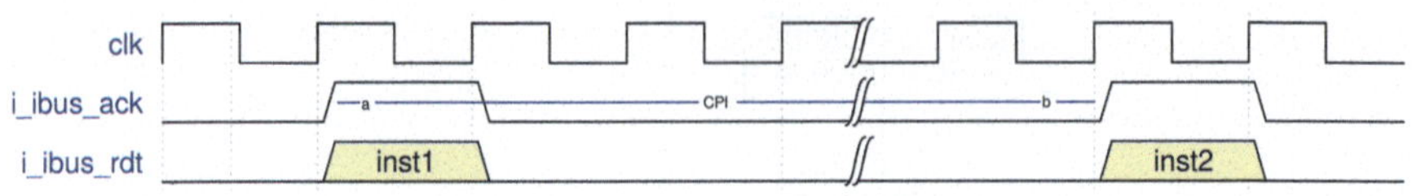

Fig. 5.10 Lifecycle of SERV instructions

between those two events (where *time*(*e*) is the number of clock cycles since the start until *e* happened). We call this difference Cycles per Instruction (CPI). This is shown in Fig. 5.10 where the first instruction starts at time a. At time point b the next instruction starts, thus the CPI for this instruction can be calculated by the difference between b and a.

CPI Analysis The WAWK program for the CPI analysis is shown in Listing 5.18. The analysis program consists of four statements. First, after the program is started and before the waveform analysis is kicked off, the first statement (Line 1) is executed once. This is triggered by the special `BEGIN` condition. This program uses the *riscv-model* Python package to decode RISC-V instructions. Therefore, an external Python script containing a few lines of Python glue code is imported (Line 2), and the list of results which will store all measured CPI values is initialized. Additionally, some aliases are created for often used signals. Next, some aliases are declared which can be used instead of the long full signal names.

The second statement (Line 9) is executed at the last cycle of each instruction. However, the statement is only executed when the opcode of the current instruction matches the opcode which we want to analyze. This opcode has to be passed to the program as a command line argument. It is available to the WAWK program in the `args` variable. At this point, the difference to the starting time of this instruction is calculated and added to the list of measured CPI values.

The next statement (Line 13) is executed whenever a new instruction starts executing. At this moment, the current time index, which is used to calculate the runtime of this instruction in the previous statement, is stored, and the current opcode is decoded and stored in the `op` variable.

Finally, after all indices were processed by the program, the final results, which consist of the average, minimum, and maximum runtimes, are printed by the last statement (Line 18).

```
1  BEGIN: {
2    import(extern);
3    cpis = [];
4    alias(clk, "TOP.servant_sim.dut.cpu.clk");
5    alias(fire, "TOP.servant_sim.dut.cpu.i_ibus_ack");
6    alias(instruction, "TOP.servant_sim.dut.cpu.i_ibus_rdt");
7  }
8
9  clk, !fire, fire@2, op == args[0]: {
10   cpis = cpis + ((INDEX - start) / 2);
11 }
```

```

clk, fire: {
  start = INDEX;
  op = call(extern.decode, instruction);
}

END: {
  if (cpis) {
    if (min(cpis) == max(cpis)) {
      printf("%10s %10d\n", args[0], average(cpis));
    } else {
      printf("%10s %10d %10d %10d\n", args[0], average(cpis),
     min(cpis), max(cpis));
    }
  }
}
```

Listing 5.18 WAWK code for CPI the analysis

Table 5.3 shows the analysis results for each of the instructions of the SERV core. The first column lists the name of the instruction. The second lists the average number of cycles this instruction requires to be executed. The last two columns show the minimum and maximum number of instructions required to execute this instruction, respectively. The table shows that the instructions mainly fall into two categories. One half of the instructions always finishes execution in constant time, while the over half requires a variable number of cycles.

5.6 Summary

In this chapter, we proposed WAL, a novel domain-specific language for nontrivial automated waveform analysis. WAL is specifically designed for waveform analysis, featuring a wide set of functionalities that allow solving waveform analysis problems in a natural and expressive way. With a macro system, a standard library of commonly used functions, and the flexibility of the S-expression syntax, it is straightforward to adapt WAL to many use cases. We have demonstrated the capabilities of WAL for design understanding and debugging in four case studies: First, we analyzed performance metrics of several RISC-V processors. Next, we profiled a RISC-V binary on a waveform, followed by a presentation about how APB bus interfaces can be analyzed using a flexible and reusable monitoring function. Then, we presented how WAL can be extended with an embedded DSL by presenting an SQL inspired query language. Finally, we introduced WAWK, a AWK-inspired front end to WAL.

Currently, WAL is not well integrated with established verification and analysis frameworks such as UVM. Utilizing WAL thus requires learning a significantly different new language to either SystemVerilog or VHDL. However, in future work, we plan to develop a SystemVerilog bridge which will allow verification engineers to integrate WAL into existing or new verification setups.

Table 5.3 Analyzed CPI values for SERV instructions when running RISC-V architectural tests

Instruction	Avg.	Min	Max
auipc	35		
lui	35		
add(i)	35		
sub	35		
and(i)	35		
or(i)	35		
xor(i)	35		
ecall	35		
slt(i)	68		
sltu	68		
sltiu	68		
sll(i)	68		
sra	75	68	99
srai	70	68	99
srl(i)	75	68	99
jal	68	68	70
jalr	69	68	70
beq	68	68	70
bge	69	68	70
bgeu	69	68	70
blt	68	68	70
bltu	69	68	70
bne	68	68	70
lb	69		
lh	69	69	70
lhu	69	69	70
lw	69	69	70
sh	69	69	70
sw	69	69	70

Further, the current WAL reference implementation is not targeted at production size workloads. Developing an efficient interpreter targeted at waveform analysis opens up a range of interesting research questions, including optimizations and data structures, which we will address in future work.

In the next chapters, we present how WAL can be used in applications ranging from processor analysis (Chap. 6) to embedding WAL into an HDL (Chap. 7) to debugging (Chap. 8).

References

1. B. Bailey, Can debug be tamed? https://semiengineering.com/bigger-debug-challenges-ahead 2019.

2. J. Bergeron, *Writing Testbenches Using SystemVerilog*. (Springer, Berlin, 2006)
3. A. Bybell, Implementation of an efficient method for digital waveform compression. https://gtkwave.sourceforge.net/gtkwave.pdf. Accessed: 18 April 2024
4. W. Chen, S. Ray, J. Bhadra, M. Abadir, L.-C. Wang, Challenges and trends in modern soc design verification. IEEE Design Test **34**(5), 7–22 (2017)
5. Common Lisp Hyperspec. http://www.lispworks.com/documentation/lw50/CLHS/Front/Contents.htm. Accessed: 18 April 2024
6. F. Ficarelli, A. Bartolini, E. Parisi, F. Beneventi, F. Barchi, D. Gregori, F. Magugliani, M. Cicala, C. Gianfreda, D. Cesarini, A. Acquaviva, L. Benini, Meet monte cimone: Exploring risc-v high performance compute clusters, in *Computing Frontiers* (2022), pp. 207–208
7. H.D. Foster, The 2020 Wilson research group functional verification study. https://blogs.sw.siemens.com/verificationhorizons/2021/01/06/part-8-the-2020-wilson-research-group-functional-verification-study
8. H.D. Foster, A.C. Krolnik, D.J. Lacey, *Assertion-Based Design* (Springer Science & Business Media, Cham, 2004)
9. M. Fowler, *Domain Specific Languages* (Addison-Wesley Professional, Boston, 2010)
10. A. Gascón, P. Subramanyan, B. Dutertre, A. Tiwari, D. Jovanović, S. Malik, Template-based circuit understanding, in *International Conference on Formal Methods in CAD* (2014), pp. 83–90
11. J.L. Hennessy, D.A. Patterson, Understanding multidimensional verification: where functional meets non-functional. Microprocess. Microsyst. **71**, 102867 (2019)
12. IEEE Std. 1364-2001. *IEEE Standard Verilog Hardware Description Language* (2001)
13. IEEE Std. 1800.2-2020. *IEEE Standard for Universal Verification Methodology Language Reference Manual* (2020)
14. L. Klemmer, D. Große, WAL: A novel waveform analysis language for advanced design understanding and debugging, in *ASP Design Automation Conference* (2022), pp. 358–364
15. L. Klemmer, D. Große, A DSL for visualizing pipelines: A RISC-V case study, in *RISC-V Summit Europe* (2023).
16. L. Klemmer, D. Große, Programming language assisted waveform analysis: A case study on the instruction performance of SERV, in *Workshop on Open-Source Design Automation* (2023)
17. L. Klemmer, D. Große, WAVING goodbye to manual waveform analysis in HDL design with WAL. IEEE Trans. Comput. Aided Design Circ. Syst. **43**(10), 3198–3211 (2024)
18. L. Klemmer, F. Skarman, O. Gustafsson, D. Große, Surfer: A waveform viewer as dynamic as risc-v, in *RISC-V Summit Europe* (2024)
19. A. Mahzoon, D. Große, R. Drechsler, RevSCA: Using reverse engineering to bring light into backward rewriting for big and dirty multipliers, in *Design Automation Conference* (2019), pp. 185:1–185:6
20. J. Malburg, A. Finder, G. Fey, A simulation-based approach for automated feature localization. IEEE Trans. Comput. Aided Design Circ. Syst. **33**(12), 1886–1899 (2014)
21. J. McCarthy, Recursive functions of symbolic expressions and their computation by machine, Part I. Commun. ACM **3**(4), 184–195 (1960)
22. A.B. Mehta, *SystemVerilog Assertions and Functional Coverage* (Springer, Berlin, 2019)
23. M. Michael, D. Große, R. Drechsler, Localizing features of ESL models for design understanding, in *Forum on Specification and Design Languages* (2012), pp. 120–125
24. R7rs Scheme. https://small.r7rs.org/. Accessed: 18 April 2024
25. Racket Language Extensions. https://docs.racket-lang.org/guide/hash-languages.html. Accessed: 18 April 2024
26. S. Ray, I.G. Harris, G. Fey, M. Soeken, Multilevel design understanding: From specification to logic, in *International Conference on Computer-Aided Design* (2016)
27. D.H. Sarah Harris, *Digital Design and Computer Architecture, RISC-V Edition* (Morgan Kaufmann, Burlington, 2021)
28. L. Schammer, J. Runge, P. Klimach, G. Fey, Design understanding: Identifying instruction pipelines in hardware designs, in *International Conference on Modern Circuits and Systems Technologies* (2022), pp. 1–6

29. SERV. https://github.com/olofk/serv. Accessed: 18 April 2024
30. F. Skarman, O. Gustafsson, L. Klemmer, Surfer 0.1.0 (2024)
31. M. Soeken, B. Sterin, R. Drechsler, R.K. Brayton, Reverse engineering with simulation graphs, in *International Conference on Formal Methods in CAD* (2015), pp. 152–159
32. S. Vasudevan, D. Sheridan, S.J. Patel, D. Tcheng, W. Tuohy, D.R. Johnson, Goldmine: Automatic assertion generation using data mining and static analysis, in *Design, Automation and Test in Europe* (2010), pp. 626–629
33. VexRiscv: A FPGA friendly 32 bit RISC-V CPU implementation. https://github.com/SpinalHDL/VexRiscv. Accessed: 18 April 2024
34. Vision Five 2. https://www.starfivetech.com/en/site/boards. Accessed: 18 April 2024
35. A. Walsemann, M. Karagounis, A. Stanitzki, D. Tutsch, Strv—a radiation hard risc-v microprocessor for high-energy physics applications. J. Instrument. **18**(02), C02032 (2023)
36. A. Waterman, K. Asanović, *The RISC-V Instruction Set Manual; Volume I: Unprivileged ISA* (SiFive Inc. and CS Division, EECS Department, University of California, Berkeley, 2019)
37. J. Zhao, I.G. Harris, Automatic assertion generation from natural language specifications using subtree analysis, in *Design, Automation and Test in Europe* (2019), pp. 598–601

Chapter 6
Programmable Analysis of RISC-V Processors Using WAL

This chapter presents techniques for the programmable analysis of RISC-V processors using WAL. The focus on the presented processor analysis techniques lies on two central components of modern systems. First, the analysis of Central Processing Unit (CPU) performance in terms of including an analysis of the processor pipeline is presented. Then, an extensible, automated, and flexible methodology for analyzing cache performance is introduced. Both approaches are evaluated in multiple experiments on several RISC-V processors and core configurations.

6.1 Introduction

This chapter presents techniques for the programmable analysis of RISC-V processors using WAL. Thoroughly analyzing the performance of processors before they are shipped to customers is essential in guaranteeing that all requirements are met. However, processor performance analysis is challenging due to the complex nature of processor and System on a Chip (SoC) design. In particular, most processors have an intricate pipeline, multiple layers of caches, or an extensive bus network. With the rise of RISC-V, this problem has become even harder due to the modular and extensible nature of RISC-V. Today, highly application-specific SoCs can be generated using one of the many available open-source and proprietary frameworks [4, 7, 17]. They allow the composition of a custom SoC from a large set of available components. Further, many are built around HDL code generators [4, 6] or straightforward extensibility [17]. Often, these frameworks typically allow significant changes to the micro-architecture. Providing so many configuration options leads to a very large number of possible configurations, and extra care is required to verify that a configuration option has no unintended side effects. Likewise, a highly configurable design complicates the analysis, since it has to be able to react to a changing design.

L. Klemmer, D. Große, *Versatile Hardware Analysis Techniques*,
https://doi.org/10.1007/978-3-031-83093-8_6

WAL [21], the waveform analysis language presented in the previous chapter, provides a powerful foundation for writing flexible analysis programs for processor analysis. In the following sections, we present generic and programmable analysis methodologies for two important analysis tasks.

First, in Sect. 6.2, we present how CPU performance metrics can be analyzed on a simulation waveform. These include metrics such as the Instructions per Cycle (IPC) value or an analysis of the pipeline behavior. To make our presented analysis applicable to a wide variety of designs, it is based on a design-independent core algorithm and requires only little amounts of design-specific code. The methods presented in Sect. 6.2 have been published in [20] and in [22].

In Sect. 6.3, we present a cache analysis methodology based on WAL. Apart from the raw performance of a CPU, the performance of a SoC is to a large part determined by the performance of memory operations. Here, caches play an import role in hiding the high latency of reading or writing to a memory. Therefore, evaluating caches is another central analysis task. Like the CPU analysis, this methodology is design-independent and thus applicable to a wide variety of caches. This section is based on a conference paper which is currently under review.

Finally, in Sect. 6.4, we summarize the presented processor analysis techniques.

6.2 RISC-V CPU Performance Analysis

Since RISC-V introduction, we have seen an explosion of, often freely, available RISC-V cores. However, this development brings its set of challenges. The huge number of available RISC-V cores, which are frequently highly configurable and extensible, makes it very hard and time-consuming for both, designers and users, to compare different cores against each other [13, 24]. One method of comparing different cores is benchmark applications that perform a set of representative computing tasks (see, e.g., [14] or [1]). They measure the run time of these tasks and typically give a score that indicates how well a core handled them. Yet, these benchmarks are SW applications that can only measure the performance of the core indirectly, for example, via the run time, without the possibility of directly accessing the micro-architectural state of the core. RISC-V performance counters offer another way to acquire performance information directly from the core. However, not every core implements the same counters, and it is difficult to add counters since this requires changes to the microarchitecture of the core itself.

When it comes to performance analysis, waveforms can be considered the ground truth since all metrics of either SW benchmarks or performance counters can be recreated from the waveform data. The waveform captures all micro-architectural information of the processor over the complete run time of the executed application. Based on this comprehensive data, arbitrarily complex performance metrics can be constructed that can measure the performance of each component of the core in detail (e.g., the pipeline, caches, or the bus interface).

```
(defun calc-ipc [traces]
  (for [trace traces]
    ;; load the trace
    (load trace 't)
    ;; if defined, run setup function
    (when (defined? 'setup) (setup))
    ;; count num of executed instructions and ipc value
    (let ([instructions (count (&& (is-valid) (instr-done)))]
          [ipc (/ 1 (/ (count (is-valid)) instructions))]])
      ;; print results
      (printf "%40s: %15.2f\n" trace ipc)
      ;; if defined, run clean-up function
      (when (defined? 'clean-up) (clean-up))
      ;; unload the trace
      (unload 't))))
```

Listing 6.1 Generic WAL function for the IPC analysis

```
(defun is-valid []
  (&& (rising TOP.IO_CLK) TOP.IO_RST_N))

(defun instr-done []
  TOP.ibex_simple_system.u_top.u_ibex_top.u_ibex_core.
  \textcolor{red}{{\protect\hookrightarrow}} id_stage_i.
    instr_done)
```

Listing 6.2 IBEX specific code for the IPC analysis

In this section, we demonstrate the use of WAL to analyze performance metrics of various RISC-V cores. These cores range from extremely area efficient ones [3] to pipelined cores with higher performance and many configuration options [4, 12]. With WAL, it is possible to split the analysis problem into a general and a core-specific part. Thus, we can write the code in such a way that the core analysis logic is independent of any core-specific information. In this section, we present a generic WAL riscv-library, thanks to which the analysis and comparison of various significantly different RISC-V cores are possible with just a few lines of processor-specific glue code.

6.2.1 Instructions per Cycle (IPC)

First, we analyze the raw performance of each core in terms of executed IPC. Since all analyzed cores are single-core architectures, the best theoretical IPC score in this case study is 1.0 instructions per cycle. This means that the core executes and commits one instruction in each clock cycle. However, this is almost impossible to achieve, for example, due to branching and memory induced delays.

The WAL program for IPC analysis is split into two separate parts, a generic and core-independent analysis part and the core-specific code, which has to be provided by the user. Listing 6.1 shows the code for the generic IPC analysis program in WAL. This function performs the IPC analysis for all waveforms passed in the *traces* parameter. For each trace, first, the trace is loaded on Line 4 and then the optional *setup* function is called on Line 6. The optional *setup* and *cleanup* functions can be defined by the users to perform core-specific setup and clean operations. Then, the number of executed instructions is calculated on Line 8 using the user-supplied *is-valid* and *instr-done* functions (see below). The idea is to count how often the predicates *is-valid* and *instr-done* evaluate to 1 on the waveform. Next, the resulting IPC value is calculated on Line 9. We divide the number of total valid cycles by the number of executed *instructions*, take the reciprocal value, and print it on Line 11. Finally, the optional *clean-up* function is called, and the trace is unloaded from the WAL environment on Line 15.

To perform the IPC analysis on a new RISC-V core, users only have to provide the two *is-valid* and *instr-done* functions. Lines 2–6 in Listing 6.2 show the implementations of these functions for the IBEX processor [12]. The IBEX processor always sets the *instr_done* signal inside the *id_stage_i* module to 1 whenever an instruction is completed. Therefore, the *instr-done* function only has to return the value of this signal. The IBEX core executes instructions when the clock is rising and reset is low, which is checked in the *instr-valid* function.

6.2.2 *Pipeline Stall Analysis*

Our WAL riscv-library also provides a function to calculate the relative number of cycles at which at least one stage of the pipeline was stalled. This metric is useful, for example, to assess how efficient the branch prediction is working. Similar to the IPC analysis, the pipeline stall function is also written in a generic way such that new cores can be easily analyzed by providing certain functions. Listing 6.3 shows the implementations for the required `is-stalled` and `is-valid` functions for the VexRiscv core. On Line 1, we first create a variable that will be used to store the names of all pipeline stages that were found in the core. This variable will be updated by the `setup` function (Line 3) whenever a new trace is loaded. The number of pipeline stages of a VexRiscv configuration pipeline can be changed. However, our `setup` function automatically works for every number of pipeline stages, since each stage contains an *isMoving* signal.

The `is-stalled` function (Line 6 in Listing 6.3) should return true whenever at least one stage of the pipeline is stalled. To check this, we create a list of the states of each pipeline stage by iterating over all stages and getting the signal value of the *isMoving* signal. If this list contains a 0, we know that one of the stages is stalled, since it is not moving forward. Finally, the `is-valid` function (Line 10) should return true at each valid cycle, i.e., a rising clock edge and no reset.

```
(define stages '())

(defun setup []
  (set! stages (groups "isMoving")))

(defun is-stalled []
  (in 0 (for/list [stage stages]
          (in-group stage #isMoving))))

(defun is-valid []
  (&& (rising TOP.clk) (! TOP.reset)))
```

Listing 6.3 VexRiscv specific code for the pipeline stall analysis

6.2.3 IPC and Pipeline Analysis Results

Table 6.1 lists the analysis results for multiple open-source cores and one commercial RISC-V core (*TGC*). The name of the core is shown in the first column, and the analyzed configuration is shown in the second. The last two columns contain the analysis results. It can be seen that *TGC*, even in the medium-sized 3-Stage configurations, achieves the second-best IPC value of all cores. Furthermore, even the small *TGC's* 3-Stage configuration achieves competitive IPC values. *IBEX* has

Table 6.1 Analysis results

Core	Configuration	IPC	Stalled Cycles
SERV	servant	0.02	*not pipelined*
PicoRv32	default	0.24	*not pipelined*
VexRiscv	microNoCsr	0.33	63%
VexRiscv	smallest	0.33	66%
VexRiscv	smallAndProductive	0.42	54%
VexRiscv	smallAndProductiveICache	0.47	51%
VexRiscv	twoThreeStage	0.47	48%
VexRiscv	secure	0.57	42%
VexRiscv	linux	0.59	38%
VexRiscv	full	0.57	35%
VexRiscv	fullNoMmuMaxPerf	0.63	33%
IBEX	default	0.63	48%
IBEX	icache	0.89	19%
TGC	3-Stage	0.61	65%
TGC	4-Stage v1	0.72	49%
TGC	4-Stage v2	0.70	45%
TGC	4-Stage v3	0.70	44%
TGC	4-Stage v4	0.68	43%
TGC	5-Stage	0.78	40%

to be set to a large cache-enabled configuration to achieve better performance. Additionally, *TGC* also achieves one of the lowest values for the relative number of cycles in which at least one of the pipeline stages is stalled.

With our generic WAL riscv-library, we were able to analyze two detailed performance metrics of cores, ranging from extremely small bit-serial cores all the way to highly configurable commercial cores [2]. Further, only a few lines of code are required for each core to be made compatible with our WAL riscv-library, and, if written accordingly, this glue code is applicable even to very different configurations of the same core.

6.2.4 Execution Time Analysis of Single Instructions

The highly modular nature of RISC-V is reflected in the extremely high configurability of many RISC-V processors. Many of the available cores today can not only be configured to support a wide range of RISC-V extensions, but they also allow a mix and match of nearly every piece of the processor, including but not limited to arithmetic components, the number of pipeline stages, different branch prediction implementations, caches, and even to entirely different bus architectures. Configuring a parameter of a processor can have a significant impact on the performance of the core. Developers might be able to make a rough estimate of this impact. Nonetheless, considering all possible ways these changes can impact the pipeline, reliable measurements are needed. For this, only an extensive simulation at the RTL level can be the ground truth.

We developed an additional analysis program to get a more accurate understanding of the performance impact of various configurations of the TGC processor. This program computes the time each instruction type (e.g., *add*, *mul*, or *srli*) takes on average to be executed. With it, we analyzed performance critical instructions for the six configurations of the *TGC* processor. Especially arithmetic instructions are suitable for optimization, often at the cost of increased hardware size. In this section, we analyze two arithmetic instructions, *div* and *mul*, from the RISC-V M extension and two shift instructions, *slli* and *srli*.

The *run* function which computes the execution time of instructions is shown in Listing 6.4. The general idea behind this function is that instructions are tracked, while they flow through the pipeline of the processor. Whenever a new instruction enters the pipeline in the first stage (called *entry* stage in the following), the current simulation time is stored in an array (i.e., a hash map). Then, when the instruction leaves the last pipeline stage (called *exit* stage in the following), the execution time is determined using the previously stored start time. Still, since instructions can be removed from the pipeline at any stage, we also have to delete instructions from the array if they are removed.

Since the number of pipeline stages varies, the function shown in Listing 6.4 is written in a generic way. The user has to supply the name of the trace file and the names of the first and last stages to the function (Line 3). Then, the trace

```
(import extern)

(defun run [trace entry-group exit-group]
 (load trace 't)
 (print TRACE-FILE)

 (define alive (array))
 (define stats (array))
 (define clk-group
   (first (groups clk reset lastStageIsFiring)))
 (define
   stages (groups arbitration_removeIt INSTRUCTION))

 (whenever (in-group clk-group (&& (rising #clk) (! #reset)))
   (in-groups stages
     (let ([instr #INSTRUCTION]
           [valid #arbitration_isValid])
       (cond
         ;; detect instructions entering pipeline
         [(&& (= CG entry-group)
              (unstable #INSTRUCTION)
              valid
              (! (in instr alive)))
          (seta alive instr INDEX)]
         ;; detect instructions exiting pipeline
         [(&& (= CG exit-group)
              #arbitration_isFiring
              valid
              (in instr alive))
          (let ([t (/ (- INDEX (geta alive instr)) 2)]
                [opcode (call extern.decode_opcode instr)])
               (dela alive instr)
               (seta stats
               (if (= opcode "invalid")
                   (+ (convert/bin instr[31:25] 7)
                      " "
                      (convert/bin instr[14:12] 3)
                      " "
                      (convert/bin instr[6:0] 7))
                   opcode)
               (+ (geta/default stats '() opcode) t)))])

       (when (&& #arbitration_removeIt (in instr alive))
         (dela alive instr)))))

 (mapa (fn [opcode ts]
           (printf "%-22s %8d %8d\n"
                   opcode
                   (average ts)
                   (length ts)))
       stats)
 (unload 't))
```

Listing 6.4 Instruction Execution Time Analysis for TGC

file is loaded (Line 4). Subsequently, the arrays for tracking the currently living (i.e., executing inside the pipeline) instructions (*alive*) and the final results (*stats*) are initialized (Lines 7–8). Additionally, the location of the clock is identified (Line 10) and groups of the pipeline stages (each has an *arbitration_removeIt* and an *INSTRUCTION* signal) are searched for (Line 12).

The main loop of the function is evaluated at each index at which the *clk* is high and *reset* is low (Line 14). Now, the function goes through each pipeline stage (Line 15) and checks if it is the entry (starting on Line 19) or exit stage (starting on Line 25). Then, the current instruction and the validity of this instruction are stored (Lines 16–17), since they are required multiple times in later lines.

If a new valid instruction enters the entry stage, it is added to the array of currently living instructions (Line 24). If an instruction is executed in the exit stage, the execution time is calculated based on the stored start time of this instruction (Line 30). Then, the opcode of the instruction is determined with the help of a call to the "riscv-model" Python package (Line 31). The instruction is removed from the array of currently alive instructions (Line 32), and the result is stored to the list in the stats array (Lines 33–41). If the RISC-V core removes an instruction from the pipeline (e.g., in case of a wrongly taken branch), we also have to remove it from the alive array (Lines 43–44). Please note that this scenario is possible within each pipeline stage. Finally, the results of the analysis are printed (Lines 46–50), and the trace is unloaded from WAL (Line 52).

Table 6.2 shows the resulting execution times for the four analyzed instructions in number of cycles. All waveforms were generated by running the Dhrystone benchmark on the specified core configurations. The first column gives the instruction name, while the remaining columns show the processor analyzed configuration. The data shows that the runtime of the *div* and *mul* instructions is significantly lower in the configurations *3-Stage* and *4-Stage v1*. The same also applies for the runtime of the shift instructions *slli* and *srli* which is much slower in the configurations *4-Stage v2* and *4-Stage v4*. With these fine-grained measurements of the hardware cost associated with each configuration, developers can decide which configuration makes the most sense for a given application and its requirements.

Table 6.2 Average execution time of instructions (in number of cycles)

Instruction	3-Stage	4-Stage v1	4-Stage v2	4-Stage v3	4-Stage v4
div	6	6	68	68	14
mul	4	4	62	62	7
slli	1	2	12	4	12
srli	1	2	21	5	25

6.3 Cache Analysis Problem

With growing demands on computing performance, industry and academia are shifting to application-specific designs [18]. This shift is complemented by the emergence of a range of sophisticated new HDLs and hardware generators. Both facilitate the customization of a hardware design to meet application-specific requirements by allowing to easily generate thousands of combinations of Intellectual Property (IP) and their parameters. For example, in many SoC designs, that are leveraging new HDLs and hardware generators [19, 27], the modification of a single line suffices to switch the bus implementation, such as transitioning from AXI to Wishbone. Likewise, changing complete components and behaviors (e.g., peripherals, branch prediction, or caches) can be as simple as adding or removing a few lines in a configuration file.

As the disparity in speed between processors and main memory widened, *caches* became very crucial hardware components. Caches provide faster access to frequently used data and instructions, reducing the time processors spend waiting for information from slower main memory. They exploit the principles of temporal and spatial locality to store relevant data, optimizing data retrieval efficiency. Evaluating cache performance is paramount for system design due to its direct influence on memory hierarchy efficiency. A meticulous analysis of cache behavior enables architects to fine-tune memory access patterns and enhance overall computational performance. Thus, many approaches have been developed, including, for instance, static cache analysis, cache simulation, analytical cache models, and utilization of performance counters to provide guidance in high-level architecture decisions. Important cache parameters strongly connected to the cache architecture include in particular cache size, type (e.g., direct mapped, set associative, and fully associative), replacement policy, and writing strategies. However, when it comes to implementing a cache in an HDL, the cache performance has to be analyzed also at this level. Indeed, all the abovementioned approaches do not target the HDL design process. In fact, evaluating major performance metrics (such as cache hits, cache misses, conflict misses, ...) as well as relating the results to software is left to the HDL designer. This means that they have to set up complex testbenches or extend the cache design with performance counters and "export" this information as the basis for the metric computation. This, however, is clearly not efficient and requires significant effort, in particular HDL code extensions as well as running simulations over and over again.

In this section, we present an *extensible, automated, and flexible methodology for analyzing cache performance of hardware designs*. Our methodology eliminates the need for creating cache models and requires no in-depth knowledge about cache internals, as it derives all information by monitoring the cache's interfaces. It is based on Waveform Analysis Language (WAL), leveraging WAL's strong support for writing generic and flexible programs. These generic programs then require only very little "glue code" to bind them to concrete hardware designs, which makes them very flexible and widely reusable. The proposed cache analysis methodology consists of three components that leverage WAL to be as flexible as possible: The main component is the *generic cache analysis algorithm* that detects requests to the cache, responses, and request drops. This algorithm does not analyze any metric and is only detecting and forwarding events to the concrete analysis metrics.

Those *analysis metrics* are the next component of our methodology. Analysis metric describe either a single value (e.g., hit-rate, average delay) or a data aggregation function (e.g., collect the moving average during simulation to generate a chart). Each metric can define fully encapsulated local variables and can implement a few callback functions that are invoked by the main algorithm. All metrics are executed separately, don't interfere with others, and are defined using convenient macros. Our methodology contains a set of predefined metrics, but adding new ones is also straightforward and requires no changes to the main algorithm.

The final component of the proposed methodology is the *glue code* that is used by the generic algorithm to detect events such as requests or responses. With much of the analysis abstracted away, this glue code usually consists of only a few lines of code per design, which specifies the involved signals or simple conditions on those signals.

We demonstrate that our methodology is applicable throughout all stages of the hardware development cycle from the first test, over debugging, all the way to analyzing multimillion cycle simulations. However, since our methodology does not depend on where the waveforms come from, it can also be applied in a post-silicon setting, for example, with waveforms produced by logic analyzers. Regardless of the analysis' scope, the generic algorithm of our methodology typically remains the same, i.e., in rare cases, only minimal adjustments are necessary. Our approach allows analyzing more than just the most basic metrics such as hit/miss rates. For example, it makes it is possible to visualize cache behavior or to combine the cache performance analysis with the software that is exercising the caches.

Numerous approaches have been developed to perform cache analysis from various perspectives. In essence, several directions can be distinguished.

Static cache analysis refers to a method of evaluating cache behavior without executing the actual program [5]. It involves inspecting the program's source code, intermediate representations, or binaries to make predictions about cache performance, leveraging abstract interpretation to derive the worst-case execution time and safe approximations [29]. As designers are interested in the average case execution behavior, combinations with symbolic execution have been developed, e.g., [11].

To evaluate new designs and to address the increasing design complexity, cache simulators have been proposed. A survey describing 28 cache simulators can be found in [9]. Cache simulators come in various types tailored for different analysis needs. Trace-driven simulators use prerecorded memory access traces for evaluating cache behavior in specific program executions [28]. Execution-driven simulators simulate cache interactions in real time during the execution of actual programs on the host platform, providing insights into dynamic scenarios. Full-system simulators simulate entire computer systems, including processors and peripherals, providing a comprehensive environment for evaluating cache behavior in a holistic system context (e.g., [8]).

Analytical simulators employ mathematical models to predict cache behavior theoretically, offering valuable insights without the need for program execution (see e.g., [15, 16]).

Finally, there are approaches leveraging performance counters (or more general performance monitoring units) [26] as well as respective APIs to access these hardware units [23]. In the same line of research, approaches using vendor-specific interfaces can be utilized to perform live cache inspection on hardware [25].

In contrast to the reviewed approaches, our proposed methodology is complementary as it focuses on the support of the HDL design process. Our goal is to provide a generic reusable solution which allows computing cache performance metrics on waveforms without the need to extend the HDL (e.g., with performance counters etc.) or complex testbench infrastructure.

This section is structured as follows. First, we introduce our cache analysis methodology in Sect. 6.3.1. There, we describe the generic algorithm, the generic metrics, the design-specific glue code, and how the methodology can be applied to waveforms. In Sect. 6.3.6, several experiments ranging from evaluating cache configurations to the analysis of multimillion cycle waveforms are presented. Finally, this section is concluded in Sect. 6.4.

6.3.1 Cache Analysis Methodology

In this section, we present an overview of the components of our cache analysis methodology. The goal of our methodology is to provide an extensible and flexible way to perform cache analyses. In order to achieve this, our methodology is split in three components: (1) the generic analysis algorithm, (2) the analysis metric implementations, and (3) the design-specific glue code.

The first two components achieve the first goal of extensibility, since the analysis algorithm is generic and works across designs, and since new metrics can be easily added in a plug-and-play fashion without changes to the generic algorithm. Metrics can range from basic hit/miss rates to the generation of visualizations of the cache behavior over the simulation time. Of these parts all but the glue code can be completely reused across widely different designs.

An overview of all components of our methodology is shown in Fig. 6.1. All components left of the dotted line are reusable across designs, and all design-specific components are shown on the right side of the dotted line. In the rest of this section, we present all of the three components in detail. The generic algorithm is introduced in Sects. 6.3.2 and 6.3.3 presents how performance metrics can be implemented, Sect. 6.3.4 discusses how the analysis is bound to a specific design using small amounts of glue code, and Sect. 6.3.5 concludes this section by showcasing how the analysis is applied to a waveform.

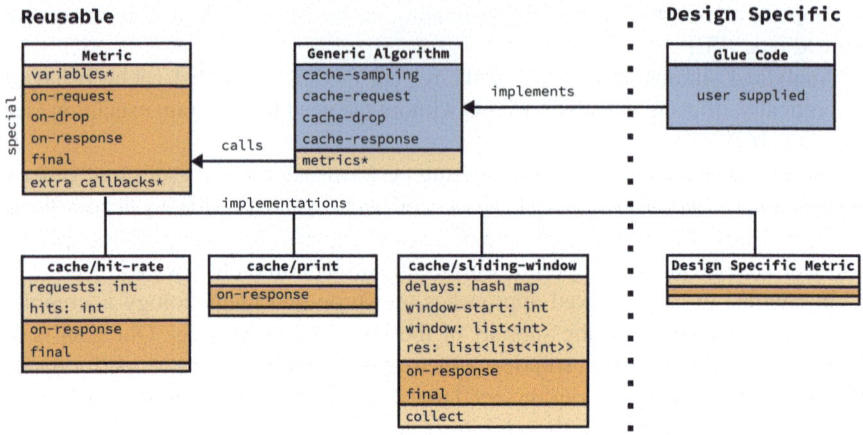

Fig. 6.1 Cache analysis methodology overview

6.3.2 Generic Cache Analysis Algorithm

The most abstract layer of our methodology is the generic cache analysis algorithm (top center in Fig. 6.1). The abstract idea of our methodology can be summarized as follows: The cache analysis runs over the waveform, trying to detect certain events (marked blue in Fig. 6.1). If an event is detected, a callback function is called for every selected metric (marked dark orange in Fig. 6.1). Callback functions then perform the computation of the metric (e.g., calculate the time between now and the last time the callback was called). They can store their (temporary) results in metric-specific variables (marked light orange in Fig. 6.1).

In more detail, this means that the first step of our cache analysis is performed by the user by selecting at which points of the waveform data is sampled. This defines the unit of a step forward or backward. For example, if no sampling condition is specified, the timepoint index is incremented whenever any signal in the waveform changes. However, if the waveform is sampled, for example, only on rising clock edges, we can calculate the delay in clock cycles by subtracting the index at which the request was detected from the index at which the response occurred.

Next, the algorithm iterates over all sampled timestamps and checks if the cache is requested using the design-specific glue code. If this is the case, the algorithm stores the current time and moves forward until one of three conditions is satisfied. The first condition is a successful response by the cache (can be either a cache hit or a miss). In this case, the algorithm calls the *on-response* callback functions of all registered analysis metrics. The second condition is satisfied when the request is dropped, for example, due to a cache flush. In this case, the algorithm calls the *on-drop* callback functions of all registered analysis metrics. Finally, the last condition is satisfied if the end of the trace is reached. In this case, no callback functions are invoked, and the request is ignored. Now, the algorithm advances to the next sampled timestamp to repeat the same process.

After the end of the waveform is reached, the finalization callback functions of all registered analysis metrics are called, and their results are collected in a list which is then returned as the result of the analysis algorithm.

6.3.3 Cache Metrics

The second most abstract layer of our methodology is the metrics. With the algorithm presented in Sect. 6.3.2, our methodology can traverse waveforms and detect events that are important for performing cache analyses. However, it does not compute any metric on its own. This task is exclusively handled by the metric implementations. By splitting the computation of metrics from the generic algorithm, it becomes easy to add or remove new metrics to the analysis without touching the main algorithm. This also allows distributing the cache analysis as a library, enabling users to easily add new (generic or design-specific) metrics. In Fig. 6.1, the generic metric is shown in the top-left corner.

New metrics can be defined using the *define-cache-metric* macro. Metrics can either be reusable across deigns (cf., *cache/hit-rate* in Fig. 6.1) or design-specific. The general usage pattern of this macro is shown in Listing 6.5. Each metric can define local variables (Line 3) and local functions (Line 5). These variables and functions live in their own namespace and are only accessible to other functions and callbacks within the same metric. This means that all metrics are separated from another and can be evaluated independently.

Next, each metric can, but does not have to, define the special callbacks shown in the middle section of the *Metric* component in Fig. 6.1 (dark orange). This is shown in Listing 6.5 on Lines 7–10. These callbacks form the core functionality required by most cache analyses. The *on-response* and *on-drop* callbacks receive as arguments the time at which the current request started, the time at which the response arrived or the drop occurred, and the upper-bound timestamp of the current analysis. This upper bound is useful for writing metrics that allow a parallel trace analysis. Finally, metrics can define extra callbacks (Line 12). These callbacks receive no arguments and get called on every sampled timestamp before the conditions for the special callbacks are evaluated.

6.3.3.1 Exemplary Metric Definition

Listing 6.6 shows the definition of the metric `cache/average-delay` which analyzes the average delay of cache responses. To perform the analysis, the metric implementation has to keep track of two values. First, the number of responses and second, the accumulated delay. Two variables are defined for this purpose on Line 2 and on Line 3. Next, the *on-response* callback function is defined on Lines 4–7.

```
(define-cache-metric name
  ;; Local Variables
  (define var 0)
  ;; Local Functions
  (defun f-internal [] ...)
  ;; Special Callbacks
  (define-callback on-request [])
  (define-callback on-response [req resp end])
  (define-callback on-drop [req drop end])
  (define-callback final [])
  ;; Extra Callbacks
  (define-callback cb1 [] ...))
```

Listing 6.5 Usage pattern for defining new analysis metrics

```
1  (define-cache-metric cache/average-delay
2    (define responses 0)
3    (define acc-delay 0)
4    (define-callback on-response [req-t resp-t end]
5      (inc responses)
6      (set! acc-delay (+ acc-delay
7                         (- resp-t req-t))))
8    (define-callback final []
9      (if responses
10         (/ acc-delay responses)
11         "no responses")))
```

Listing 6.6 Implementation of the average delay metric

Whenever it is called, the number of responses is incremented (Line 5), and the observed delay for the current request is added to the accumulated delay counter variable (Line 7). The *final* callback function is defined on Lines 8–11. If the variable *responses* is not 0, the result is the accumulated delay divided by the number of responses, else an error message is returned.

6.3.4 Design Specific Code

The final and lowest layer of our methodology is the glue code that implements the events required by the generic algorithm (c.f., Fig. 6.1) and the metrics. Since the two previous layers work on abstract events and callbacks, they can be reused across all designs. In practice, this leaves only the lowest layer with design-specific code, which often consists of only a few lines. In Fig. 6.1, all parts related to the glue code are highlighted in blue.

This section presents the code required to apply our methodology to the VexRiscv processor. VexRiscv is an open-source, pipelined, and highly parameterizable RISC-V processor. The glue code for VexRiscv's data cache is shown in Listing 6.7.

```
1  (alias clk 'dut.dataCache_1.clk)
2  (alias rst 'dut.dataCache_1.reset)
3  (alias valid 'dut.dataCache_1.io_cpu_execute_isValid)
4  (alias stuck 'dut.dataCache_1.io_cpu_execute_haltIt)
5  (alias miss 'dut.dataCache_1.io_cpu_execute_refilling)
6  (alias flush_rdy 'dut.dataCache_1.io_cpu_flush_ready)
7  (alias flush_vld 'dut.dataCache_1.io_cpu_flush_valid)
8
9  (defmacro cache-sampling []
10   '(&& (rising clk)
11        (low rst)))
12
13 (defmacro cache-request []
14   '(&& (high valid)
15        (|| (! stuck) miss)
16        (! (high valid miss))@1))
17
18 (defmacro cache-drop []
19   '(high flush_rdy flush_vld))
20
21 (defmacro cache-response []
22   '(&& (high valid) (low stuck miss)))
```

Listing 6.7 Glue code for the VexRiscv data cache

To improve readability, we first define some aliases for the long signal names on Lines 1–7. Next, we implement the required events. As previously stated, the *cache-sampling* event specifies which time points are considered by the analysis. In combination with WAL's `sample-at` function, this event can be used to practically set the unit of the delay we are analyzing. In the case of VexRiscv, we sample at every timestamp at which the clock is rising and at which the reset signal is low. Requests are dropped if the *flush_ready* and *flush_valid* signals are both high (Lines 18–19).

Next, the *cache-request* is implemented on Lines 13–16. In the case of VexRiscv, a new request starts when the *valid* signal is high and the core is not stuck due to a cache miss (Line 15). Additionally, we have to check if we are currently in an ongoing request which has seen no response until now (Line 16). Finally, the *cache-response* event is implemented on Lines 21–22.

6.3.5 Running the Analysis on a Waveform

With the processor-specific glue code defined, the analysis can be run on a waveform as shown in Listing 6.8. First, on Line 1 the cache analysis library is imported. This makes the generic algorithm and a number of predefined metrics available. Next, the waveform is loaded into WAL on Line 2 and the timestamps at which signals should be sampled are set using WAL's `sample-at` function. A list of all metrics

```
1 (use cache-analysis)
2 (load "waveform.fst" 't)
3 (sample-at (find (cache-sampling)))
4 (define callbacks
5   (list cache/hit-rate cache/average-delay))
6 (define results (analyze-cache callbacks 0 MAX-INDEX))
```

Listing 6.8 Running the analysis on a waveform

that should be analyzed is created on Line 5. Please note that *cache/hit-rate* and *cache/average-delay* metrics are included with the cache analysis library. Finally, the analysis is started by a call to the *analyze-cache* function, together with the list of callbacks and the time range that should be analyzed on Line 6. In this example, the analysis is applied on the whole waveform from the first timestamp 0 to the last timestamp, which can be read using the `MAX-INDEX` special variable available in WAL.

6.3.6 Experiments

In this section, we present experiments that present how our cache analysis methodology is applicable without changes to the automated analysis of various processor configurations, debug scenarios using visualizations, and large-scale simulation analysis using parallelization.

6.3.6.1 Automated Configuration Analysis

In this case study, we automatically analyze various configurations of the VexRiscv processor. We analyze both, data caches and instruction caches, for cache sizes ranging from 1024 bytes to 32768 bytes and with cache line sizes of 32 and 64 bytes. In the case of the data cache analysis, the instruction cache size is fixed to 8192 bytes with a line size of 32. In the case of the instruction cache analysis, the data cache size is fixed to 8192 bytes with a line size of 32.

Both caches are connected to the same memory via a Wishbone interface. To get realistic delay results in our simulation testbench, we configured the memory to have a read and write delay of 20 cycles. To exercise the caches, we ran the *nettle-aes* benchmark of the Embench suite [1].

Data Cache The results of four performance metrics of the data cache are shown in Table 6.3. The first two columns list the cache size and cache line size, respectively. The next column lists the IPC value for this simulation run. To compute the IPC value, we defined a new metric that utilizes an extra callback function to count the number of executed instructions and the number of all clock cycles over

Table 6.3 Data cache results

Size	L. Size	IPC	Delay*	Miss Delay*	H. Rate (%)
1024	32	0.052	43.297	175.418	75.317
1024	64	0.028	85.178	351.457	75.764
8192	32	0.256	6.751	174.996	96.141
8192	64	0.216	9.908	351.325	97.179
32768	32	0.380	2.011	173.346	98.840
32768	64	0.377	2.290	349.592	99.345

* In clock cycles.

```
(define-cache-metric ipc
  (define instructions 0)
  (define valid-cycles 0)
  (define-callback ipc []
    (when (instr-done)
          (inc instructions))
    (inc valid-cycles))
  (define-callback final []
    (when valid-cycles
      (/ instructions valid-cycles ))))
```
Listing 6.9 IPC Analysis Metric

the complete simulation time. The code of this IPC metric is shown in Listing 6.9. This IPC metric defines the additional extra callback function *ipc*, which is called on every sampled timestamp. It counts the number of cycles and the number of executed instruction which allows the IPC value computation in the *final* callback. We use the IPC value as a proxy to analyze how the cache impacts the overall performance of the processor. The last three columns list the average read delay over all requests in clock cycles, the average read delay of cache misses in clock cycles, and the read hit rate as a percentage.

Based on the hit rate shown in the last column, we can see that the nettle AES benchmark makes heavy use of the data caches. The two configurations with a cache size of 1024 each achieve a hit rate of 74%. Such a low hit rate combined with the high penalty of cache misses of 175 or 351 cycles leads to a very high average delay and, as a result, a low IPC value for those two configurations. With increasing cache and line sizes, the hit rate improves continuously until it is above 99%. The average delay of cache misses is coupled to the line size only, with higher line sizes resulting in a higher delay.

Instruction Cache Table 6.4 lists the performance metrics of some configurations of VexRiscv's instruction cache. Since the IPC, delay, miss delay, and hit rate metrics are generic, they can be used without change for the instruction cache as well. Only the signal names in the glue code, and some of the required event conditions had to be changed due to a slight difference in the cache's interface. In the case of the instruction cache, we varied the instruction cache size and instruction

Table 6.4 Instruction cache results

Size	L. Size	IPC	Delay*	Miss Delay*	H. Rate (%)
1024	32	0.194	1.524	201.618	0.992
1024	64	0.187	1.624	372.136	0.996
8192	32	0.256	0.196	198.179	0.999
8192	64	0.256	0.190	372.551	0.999
32768	32	0.256	0.196	198.179	0.999
32768	64	0.256	0.190	372.551	0.999

* In clock cycles.

cache line size and fixed the data cache size to 8192 bytes and a line size of 32. Compared to the data cache, even smaller instruction cache sizes achieve a much higher hit rate than the data cache. Further, we see that the IPC stops increasing (and actually is the same for data cache size 8192 as in Table 6.3) with instruction cache sizes larger than 8192. This indicates that, in this configuration, the nettle-aes benchmark is limited by the data cache.

Many components of the design influence the results from Tables 6.3 and 6.4. For example, the bus interface between the cache and the memory has an effect on the time it takes to fetch data from the memory into the cache. Due to space limitations, we omitted further configurations. However, evaluating them is only a matter of generating and simulating the configurations and running the same analysis on the resulting traces. Since the analysis is run on the waveform of the real hardware design, cycle accurate results are obtainable without having or modifying a cache model.

6.3.6.2 Visualization

Our cache analysis methodology is not only useful to analyze metrics that result in one "hard" answer, such as the hit/miss rate. Analysis metrics can also aggregate data for further processing, such as visualizations.

One example of this is shown in Fig. 6.2 where the ten functions with the most instruction cache misses of the Embench *tarfind* benchmark are shown in a bar chart. For this metric, we read the symbol table of the ELF file and created a WAL list of all functions with their names and their start and end addresses. The idea behind this metric is to check if the current response was a miss, and, if this is the case, to read the address that was requested. This address is then compared against the list of functions extracted from the ELF file, and a corresponding entry in a hash map is incremented. Finally, a Python function is called that draws the bar chart using the Matplotlib plotting library.

Another example for a visualization metric is shown in Fig. 6.3, which plots the average instruction cache delay over the course of the simulation for the IBEX [12] (formerly known as Zero-riscy) processor. The data is plotted using Matplotlib and

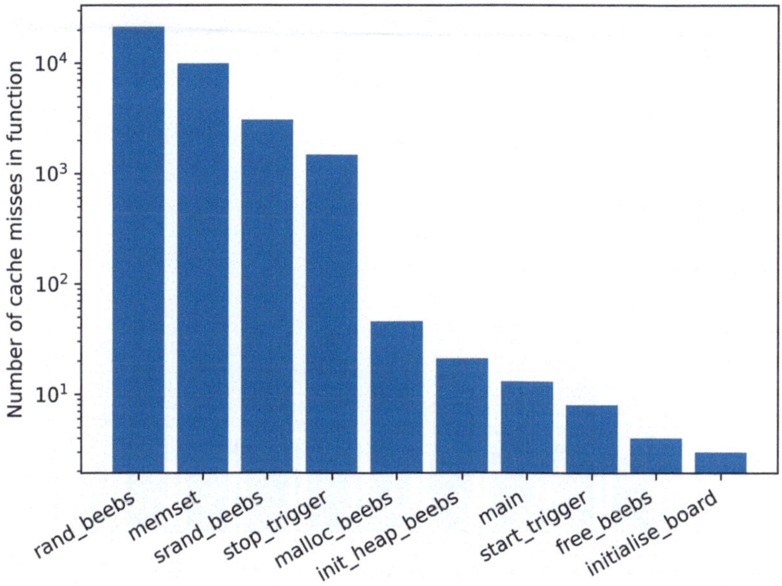

Fig. 6.2 Functions with the most instruction cache misses

aggregated by a metric that moves a sliding window (of configurable width) over the waveform.

6.3.6.3 Large-Scale Analysis

The results of simulating large designs with complex firmware are often simulation runs over millions of clock cycles. Analyzing waveforms of this size can become challenging when done in a purely sequential fashion. However, many analysis tasks can be computed on the same trace in parallel by dividing the trace into multiple slices. This is also supported by our methodology through a parameter which specifies the *end* index until which cache requests are handled. By combining this with the `step` function to set the index to some time point, we can analyze a slice of the overall waveform. Please note that the waveform is not cut at the *end* index and that requests that start within the slice but whose response occurs after the *end* index are correctly handled. In many cases, this system allows a very natural way to parallelize metrics that requires little or no changes to the normal metrics.

To show that our methodology scales to multimillion cycle simulations, we ran benchmarks for millions of cycles on two processors. We then analyzed the instruction cache of IBEX and the data cache of VexRiscv. The results of this are shown in Table 6.5. To improve the runtimes of the analysis, we parallelized it using the Cuneiform language [10] for large-scale data analysis. Using the Cuneiform, we created 64 tasks, each of which processes one slice of the waveform. After the tasks

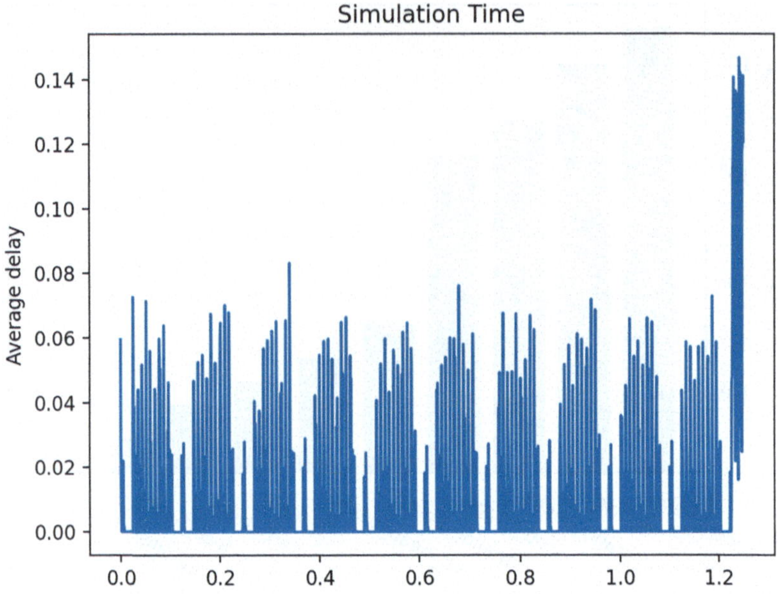

Fig. 6.3 Sliding window delay (in clock cycles) of IBEX's data cache during the simulation of the Coremark benchmark

Table 6.5 Parallel analysis results

Core	Benchmark	Cycles	Hit Rate (%)	Runtime (s)
IBEX	coremark	6,380,921	97.3	180
VexRiscv	picojpeg	4,948,761	99.6	242
VexRiscv	qrduino	5,082,492	99.9	155
VexRiscv	huffbench	7,395,706	99.8	381
VexRiscv	tarfind	16,972,672	99.9	620
VexRiscv	wikisort	18,415,326	99.9	335

are started, Cuneiform handles synchronization and the passing of all slice results to the final combination function. This function takes and combines the partial results of all slices, for example, by taking the average or the sum of all sub-results.

All results in Table 6.5 were computed on a 64-Core AMD EPYC 7713 Processor with 128 threads and 256 GB RAM. The VexRiscv configuration uses instruction and data caches of 8192 bytes, while the IBEX configuration uses an instruction cache of 1024 bytes size (IBEX has no data cache). For IBEX and VexRiscv, we used the same hit rate metric implementation as in Sect. 6.3.6.1. Additionally, we added a small WAL function that creates the start and end time points of the slices.

6.4 Summary

In this chapter, we presented techniques for programmable and extensible analysis of RISC-V processors for metrics such as instruction performance or cache performance. The presented techniques allow analyzing a wide range of different processors, since they are built around generic analysis programs that are not tied to specific designs. Thus, our proposed techniques are reusable and easily extendable to new designs. We have shown that this methodology works across all stages of the hardware development cycle in experiments ranging from automated analysis of processors using various cache configurations, over generating visual debug information to the analysis of multimillion cycle simulations. For this, the HDL designer only has to provide minimal glue code to connect their cache interface for the cache performance analysis.

References

1. Embench™: A modern embedded benchmark suite. https://www.embench.org/. Accessed: 2024-04-18
2. Minres technologies the good folks series. https://www.minres.com/products/the-good-folk-series/. Accessed: 2024-04-15
3. SERV. https://github.com/olofk/serv. Accessed: 2024-04-18
4. VexRiscv: A FPGA friendly 32 bit RISC-V CPU implementation. https://github.com/SpinalHDL/VexRiscv. Accessed: 2024-04-18
5. M. Alt, C. Ferdinand, F. Martin, R. Wilhelm, Cache behavior prediction by abstract interpretation, in *Proceedings of the Third International Symposium on Static Analysis* (1996), pp. 52–66
6. K. Asanović, R. Avizienis, J. Bachrach, S. Beamer, D. Biancolin, C. Celio, H. Cook, D. Dabbelt, J. Hauser, A. Izraelevitz, S. Karandikar, B. Keller, D. Kim, J. Koenig, Y. Lee, E. Love, M. Maas, A. Magyar, H. Mao, M. Moreto, A. Ou, D.A. Patterson, B. Richards, C. Schmidt, S. Twigg, H. Vo, A. Waterman, The rocket chip generator. Technical Report UCB/EECS-2016-17, EECS Department (University of California, Berkeley, 2016)
7. J. Balkind, M. McKeown, Y. Fu, T. Nguyen, Y. Zhou, A. Lavrov, M. Shahrad, A. Fuchs, S. Payne, X. Liang, M. Matl, D. Wentzlaff, Openpiton: an open source manycore research framework, in *Proceedings of the Twenty-First International Conference on Architectural Support for Programming Languages and Operating Systems*, ASPLOS '16 (ACM, New York, 2016), pp. 217–232
8. N. Binkert et al., The gem5 simulator. SIGARCH Comput. Archit. News **39**(2), 1–7 (2011)
9. H. Brais, R. Kalayappan, P.R. Panda, A survey of cache simulators. ACM Comput. Surv. **53**(1), 1–32 (2020)
10. J. Brandt, W. Reisig, U. Leser, Computation semantics of the functional scientific workflow language cuneiform. J. Funct. Program. **27**, e22 (2017)
11. D.-H. Chu, J. Jaffar, R. Maghareh, Precise cache timing analysis via symbolic execution, in *IEEE Symposium on Real-Time and Embedded Technology and Applications* (2016), pp. 1–12
12. P. Davide Schiavone, F. Conti, D. Rossi, M. Gautschi, A. Pullini, E. Flamand, L. Benini, Slow and steady wins the race? a comparison of ultra-low-power risc-v cores for internet-of-things applications, in *2017 27th International Symposium on Power and Timing Modeling, Optimization and Simulation* (2017), pp. 1–8

13. A. Dörflinger et al., A comparative survey of open-source application-class RISC-V processor implementations, in *Computing Frontiers Conference* (2021), pp. 12–20
14. EEMBC. Coremark. https://www.eembc.org/coremark/. Accessed: 2024-04-18
15. S. Eyerman, L. Eeckhout, T. Karkhanis, J.E. Smith, A mechanistic performance model for superscalar out-of-order processors. ACM Trans. Comput. Syst. **27**(2), 1–37 (2009)
16. A. Ghosh, T. Givargis, Cache optimization for embedded processor cores: an analytical approach, in *International Conference on Computer-Aided Design* (2003), pp. 342–347
17. C. GmbH, Codasip studio. https://codasip.com/products/codasip-studio/. Accessed: 2024-04-18.
18. J.L. Hennessy, D.A. Patterson, A new golden age for computer architecture. Commun. ACM **62**(2), 48–60 (2019)
19. M. Käyrä, T.D. Hämäläinen, A survey on system-on-a-chip design using chisel hw construction language, in *Annual Conference of the IEEE Industrial Electronics Society 2021* (2021), pp. 1–6
20. L. Klemmer, D. Große, Waveform-based performance analysis of RISC-V processors: late breaking results, in *Design Automation Conference* (2022), pp. 1404–1405
21. L. Klemmer, D. Große, WAVING goodbye to manual waveform analysis in HDL design with WAL. IEEE Trans. Comput. Aided Des. Circuits Syst. **43**(10), 3198–3211 (2024)
22. L. Klemmer, E. Jentzsch, D. Große, Programmable analysis of RISC-V processor simulations using WAL, in *Design and Verification Conference and Exhibition Europe* (2022)
23. P.J. Mucci, S. Browne, C. Deane, G. Ho, Papi: A portable interface to hardware performance counters, in *Proceedings of the Department of Defense HPCMP Users Group Conference* (1999)
24. E. Sperling, Which processor is best? https://semiengineering.com/which-processor-is-best. Accessed: 2024-04-18
25. D. Tarapore, S. Roozkhosh, S. Brzozowski, R. Mancuso, Observing the invisible: Live cache inspection for high-performance embedded systems. IEEE Trans. Comput. **71**(03), 559–572 (2022)
26. J. Treibig, G. Hager, G. Wellein, Performance patterns and hardware metrics on modern multicore processors: best practices for performance engineering, in *Proceedings of the 18th International Conference on Parallel Processing Workshops* (2012), pp. 451–460
27. L. Truong, P. Hanrahan, A golden age of hardware description languages: Applying programming language techniques to improve design productivity, in *3rd Summit on Advances in Programming Languages 2019* (2019)
28. R.A. Uhlig, T.N. Mudge, Trace-driven memory simulation: A survey. ACM Comput. Surv. **29**(2), 128–170 (1997)
29. R. Wilhelm et al., The worst-case execution-time problem—overview of methods and survey of tools. ACM Trans. Embed. Comput. Syst. **7**(3), 1–53 (2008)

Chapter 7
HDL-Integrated Waveform Analysis

This chapter presents a methodology to combine automated waveform analysis with a modern Hardware Description Language (HDL). By adding small annotations to hardware designs written in the Spade HDL, waveform analysis programs can be generated by the Spade compiler. This allows developers to not only easily analyze their own designs but also enables them to ship their designs to customers with powerful analysis programs. The usage of the presented methodology is showcased on an implementation of the Wishbone bus interface and an analysis for Finite State Machines (FSMs).

7.1 Introduction

One of the most critical aspects for the success of a new product is the time it takes from the conception of the idea to the time the product is available on the market. When it comes to shortening this *time-to-market*, no other industry has been as successful as the software industry. This agility we see in the software domain today is to a large extent enabled by continuous improvements to tools and workflows.

Compared to software tooling, the hardware domain saw relatively little advancements in this area until the emergence of the open-source EDA community, with tools such as Yosys [11, 16] or the Chisel HDL [3]. This is, in part, due to the naturally much lower abstraction level of hardware design. RTL languages, such as Verilog and VHDL, typically express very little high-level design intent to simulators, synthesis flows, and other tools. The description consists of instantiations of individual components that manipulate individual signals, but properties of those signals and instances, such as which signals make up a bus, or which registers correspond to a pipeline versus which are used for state machines, is not expressed in the language. This makes it difficult to build reusable tools for analysis and debugging, which in turn, means that debugging often has to be done through

L. Klemmer, D. Große, *Versatile Hardware Analysis Techniques*,
https://doi.org/10.1007/978-3-031-83093-8_7

manual waveform inspection, a tedious process which has to be redone for every design change.

WAL, presented in Chap. 5, is a language which allows writing programs to perform automatic waveform analysis. In the previous chapters, we presented how it can be used to analyze bus traffic, analyze the transitions of state machines, or measure the performance of processors [6, 7]. However, because RTL languages allow great flexibility in expressing designs, there is relatively little, if any, shared structure between different designs. Therefore, these analysis programs have often to be written on a per-design basis.

Spade [14] is a statically typed HDL with explicit constructs for registers, pipelines, and memories. In addition to boosting productivity by adding more abstraction, Spade exposes more high-level design intent to the compiler. Registers comprising a state machine become explicit, and pipelines being built into the language make the compiler aware of which registers are used for pipelining, and which signals logically belong together. Finally, via Spade's strict type system for defining buses and other related signal groups, such as ready-valid interfaces, information about which analysis is applicable to which signals is readily available to WAL.

By bundling WAL analysis programs with Spade projects, they can query the Spade compiler for additional information about the design, and they become automatically discoverable by the Spade toolchain. We call WAL programs bundled with Spade projects *analysis passes* as they can be automatically run as an additional step in the build and simulation flow.

The tight integration of Spade and WAL allows users to directly take advantage of these analysis passes simply by annotating the signals or structures they want the analysis to be performed on. Thanks to the knowledge the compiler has, these annotations are all that is required to connect the design with the analysis. Now, compiling, simulating, and analyzing are handled automatically by the Spade build system. In addition, this not only works for a single Spade project, but analysis passes can seamlessly be reused across the Spade ecosystem. This allows library (IP) authors to provide not only an implementation but also analysis passes, thus giving their users a head start on ensuring design quality.

The main contributions of this chapter are twofold. The first and primary contribution is the integration of WAL and Spade specifically. This includes augmenting the Spade compiler and language with additional annotations to mark types and values for analysis as well as adding several functions and macros to WAL which make it easier to write Spade specific analysis passes. In addition, we developed a plugin for the Spade build system to bundle WAL analysis passes with Spade libraries and to detect and run those automatically. This plugin is available to Spade users on GitLab.[1] The code for the example analysis passes is also available on GitLab.[2] The changes to the Spade compiler have been included in its repository.

[1] https://gitlab.com/spade-lang/wal_analysis

[2] https://gitlab.com/TheZoq2/remora-code

The second major contribution of this chapter is the methodology used to perform the integration. Its core is to bind analysis programs to designs using compiler-generated signals leveraging a strong type system. This methodology can be used to integrate WAL or similar tools with other modern HDLs.

Recently, other works have looked into increasing productivity by improving EDA tooling. There are several modern HDLs available. Perhaps the most well known is Chisel [3], a hardware construction language where hardware description takes the form of Scala programs which instantiate hardware components. Several other hardware construction languages exist, for example, SpinalHDL [15], also embedded in Scala and Amaranth [1] which is embedded in Python. Other notable projects include Clash [2], a compiler from Haskell to hardware, Pipeline-C [5], a HDL heavily inspired by C, and Silice [8], a standalone HDL primarily focused on describing algorithms. However, to the best of the authors' knowledge, none of these projects integrate the automated waveform analysis described in this chapter.

In [4], the authors explore FIRRTL [9], the Flexible Internal Representation for RTL, as a basis for hardware libraries. They also argue that traditional design languages have been slow to adopting abstraction and modularity and that these features can significantly improve the design speed and reusability. FIRRTL also provides design analysis options such as design coverage instrumentation or logic optimization. However, simulation analysis is not the main focus of FIRRTL, and the coverage analysis relies on code transformations that inject additional logic into design. Compared to [4], our approach also injects additional signals into the design. Nonetheless, these only provide the bridge to a fully fledged programming analysis that provides much more analysis capabilities. In addition, our approach emphasizes that new analysis passes are provided by library authors, and that this is possible with as little work as possible, often with just one automatically detected file inside the library. Overall, since FIRRTL also encodes the high-level design intent similar to Spade, integrating WAL analysis passes into FIRRTL should be no problem.

Yosys [16] can be used to extract FSMs from a flattened netlist after synthesis. However, this approach only provides a static view on an FSM, so state distribution and state transitions as well as frames for FSM transitions w.r.t. simulation scenarios cannot be determined.

This chapter is based on the paper published in [13] and is structured as follows. First, an example showcasing the power of the proposed automated waveform analysis is given in Sect. 7.2. Then, in Sect. 7.3, the Spade HDL is introduced. In Sect. 7.4, a more thorough example of how analysis passes are used and created is given, and Sect. 7.5 presents the built-in state machine analysis pass. Then, Sect. 7.6 presents implementation details of our proposed WAL and Spade integration. Finally, the chapter is concluded in Sect. 7.7.

7.2 Motivating Example

As a motivating example, we will study a sample design consisting of two Wishbone masters, which communicate with a shared slave via an arbiter. Wishbone [10] is a widely used on-chip communication protocol with broad adoption especially in the open-source hardware community. It utilizes handshaking-based communication and allows the implementation of various network topologies such as buses or point-to-point communication.

A Spade implementation of the Wishbone bus is available in a git repository,[3] which can be added as a dependency to a Spade project via a configuration file for the build system, Swim. This is shown in Listing 7.1.

```
[libraries.wishbone]
git = "https://gitlab.com/spade-lang/lib/fishbone"
branch = "main"
[plugins.wal_analysis]
git = "https://gitlab.com/spade-lang/wal_analysis"
branch = "main"
```

Listing 7.1 The Wishbone implementation and WAL analysis plugin can be included as a dependency in the swim project configuration file

```
entity wb_harness(clk: clock, rst: bool)
    -> (int<16>, int<16>)
{
    #[wal_trace(target=wb1, clk=clk, rst=rst))]
    let (wb1, wb1inv) = port;
    #[wal_trace(target=wb2, clk=clk, rst=rst))]
    let (wb2, wb2inv) = port;

    let _ = inst wishbone_arbiter(
            clk, rst, wb1inv, wb2inv, ...
        );
    let o1 = inst wishbone_master(
            clk, rst, wb1, ...
        );
    let o2 = inst wishbone_master(
            clk, rst, wb2, ...
        );

    (o1, o2)
}
```

Listing 7.2 Spade description of two wishbone masters communicating via a shared arbiter. The novel `#[wal_trace(...)]` annotations enable the analysis for an interface

The top level of our motivating example is shown in Listing 7.2. A more thorough description of the Spade language is given in Sect. 7.3, but for this example, the most

[3] https://gitlab.com/spade-lang/lib/fishbone

interesting statements are on Line 5 and Line 7, which define two wishbone buses, `wb1` and `wb2`. On Lines 9–17, the ports are passed to the arbiter and the two masters. The arbiter is configured to reply to port 1 with a latency of three cycles, and port 2 with a latency of six cycles (not shown in Listing 7.2). The masters perform single random reads or bursts of writes, with write bursts being more common and reads happening randomly between the write bursts.

To meet the systems requirements, passing functional tests is not enough. Let us look at the bus example: Given a workload, i.e., simulation trace, various metrics such as the delay and number of read/write transactions, average read/write latency, number of idle cycles, etc. have to be determined. Usually, the engineer starts using a waveform viewer using features like markers to determine the initial results based on fully manual inspection. However, this quickly becomes unfeasible for tasks such as finding the average delay over the complete simulation trace. Therefore, later in the design process, custom monitoring logic is added to the design, or the testbench is extended with analysis logic. Both approaches have their drawbacks as they require lots of additional work by users of the Wishbone library, either by invasive changes to the design logic for debugging purposes (which should be avoided) or by writing custom testbenches which are hardly reusable across projects.

By tightly coupling Spade and WAL, the users of the Wishbone library who wish to determine the metrics mentioned above only have to annotate the Spade code with a single additional line per Wishbone instance as shown on Lines 4 and 6 of Listing 7.2. Now, by simply running `swim plugin analysis`, the design is compiled and simulated, and the WAL analysis programs are executed automatically for every simulation trace. Listing 7.3 shows the output of the Wishbone analysis pass, which collects performance metrics. The output includes information about the number of reading and writing transactions, average read and write delays, the number of error responses by the slave, and the number of cycles in which the interface is idle. These metrics are analyzed and presented for every annotated Wishbone interface, in the case of this example the two interfaces `wb_harness.wb1` (`wb1`) and `wb_harness.wb2` (`wb2`).

```
[ANALYSIS] # wb_harness.wb1
[ANALYSIS]    Nr. transactions : 697
[ANALYSIS]    Nr. reads        : 16
[ANALYSIS]    Avg read latency : 6 clock cycles
[ANALYSIS]    Nr. writes       : 681
[ANALYSIS]    Avg write latency: 9 clock cycles
[ANALYSIS]    Inactive cycles  : 6%
[ANALYSIS] # wb_harness.wb2
[ANALYSIS]    Nr. transactions : 421
[ANALYSIS]    Nr. reads        : 0
[ANALYSIS]    Avg read latency : - clock cycles
[ANALYSIS]    Nr. writes       : 421
[ANALYSIS]    Avg write latency: 12 clock cycles
[ANALYSIS]    Inactive cycles  : 22%

```

Listing 7.3 Analysis results of the Wishbone analysis pass

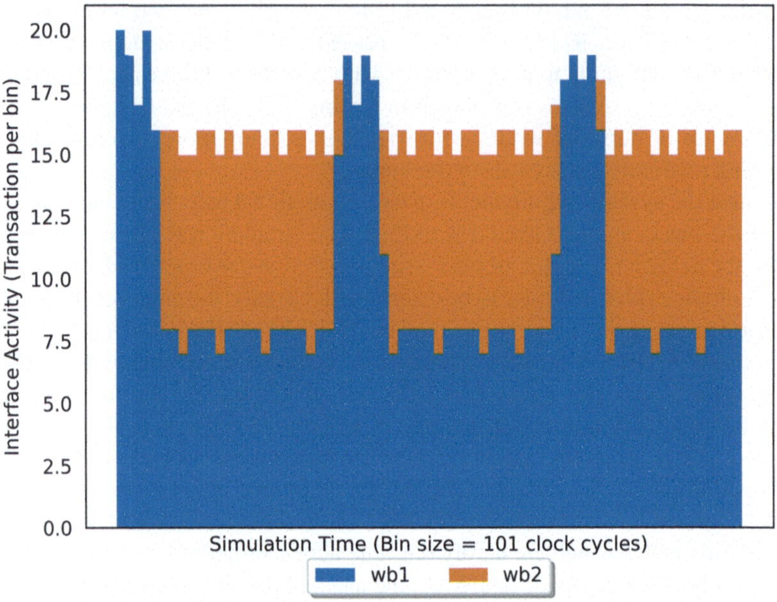

Fig. 7.1 Histogram of transactions on Wishbone interfaces. The y-axis value is the number of transactions in a given time slice of the simulation (bin)

Additionally, the Wishbone analysis pass also generates a histogram containing the transactions of every traced Wishbone interface. The generated histogram displays the activity of all interfaces over the complete simulation time using a stacked bar chart. Each colored bar of the histogram represents the number of transactions on a given Wishbone bus in a slice of the simulation time. Figure 7.1 shows the activity of the two Wishbone interfaces. From the chart, we can see that the two traced interfaces show different behaviors. `wb1` (blue) is sending a constant stream of data, while `wb2` (orange) is sending data in larger chunks. Interestingly, the number of transactions by `wb1` spikes three times. This always happens when `wb2` is not accessing the bus, thus leaving more bandwidth to `wb1`. Another notable observation is that the total number of completed transactions is higher when only `wb1` is used, compared to when both `wb1` and `wb2` are active. This is because the requests from `wb2` take longer to finish, six cycles compared to three.

From the perspective of library maintainers, providing analysis programs is as simple as placing them in an analysis directory in their libraries. Then, these programs are automatically detected by Swim and by this become analysis passes.

The key takeaway from this example is that by bundling analysis passes with libraries, and having convenient annotations for opting into those analysis passes, one can, almost without effort, get a large amount of information about the dynamic behavior of the system, information that would be very difficult to find simply by looking at waveforms in a traditional waveform viewer, or very cumbersome to generate in traditional testbenches.

7.3 The Spade HDL

In this section, we introduce the Spade HDL. Spade [14] is a statically typed open source HDL which adds abstraction without sacrificing low level control over the generated hardware. These abstractions include language level support for pipelines, explicit constructs for registers and memories, as well as support for ports with linear type checking to ensure correctness.

In addition, Spade integrates several ideas found in modern software languages. Rather than being imperative, it is expression-based, meaning that the result of a conditional, like an if expression, is assigned to a variable, instead of the variable being set in each branch. The language has type inference, meaning that types usually do not have to be explicitly spelled out, the compiler will infer them and report errors if it detects inconsistencies.

```
enum Cmd {
  Data{l: int<16>, r: int<16>}
  Mult,
  Add,
}

struct Stream<T> { valid: bool, data: T }

enum State { Add, Mult }

pipeline(3) main(
  clk: clock,
  rst: bool,
  cmd: Stream<Cmd>
) -> Stream<int<16>> {
    reg(clk) state reset(rst: State::Mult()) =
      match cmd {
        Stream(true, Cmd::Mult) => State::Mult(),
        Stream(true, Cmd::Add) => State::Add(),
        Stream(true, Cmd::Data(_, _)) => state,
        Stream(false, _) => state
      };

    let (out_valid, l, r) = match cmd {
      Stream(true, Cmd::Data(l ,r)) =>
        (true, l, r),
      _ =>
        (false, 0, 0)
    };

    let sum = trunc(l+r);
    let prod = trunc(l*r);
```

```
33     reg * 3;
34       let result = match state {
35         State::Mult => Stream(out_valid, prod),
36         State::Add => Stream(out_valid, sum)
37       }
38       result
39   }
40
```

Listing 7.4 A Spade unit processing a stream of commands into a stream of integers

For a full description of the Spade language, see [12]. However, for the purpose of this chapter, it is enough to understand the code shown in Listing 7.4, which will also be used as a running example throughout the rest of this chapter. At a high level, this defines a compute-unit which takes a stream of commands and processes them one at a time. There are two types of commands: first the mode selectors, `Mult` and `Add`, and secondly `Data` which contains two values to be multiplied or added depending on the previous mode command. The enum type containing these commands is defined on Lines 1–5. The input and outputs to the module are streamed with a data valid signal. Line 7 defines this stream type as a struct that is generic over type T and contains the valid bit as well as data of type T. The last type for this example is defined on Line 9. It specifies the internal states that the unit can have: `Add` and `Mult`.

The implementation of the unit starts on Line 11, beginning with its name and specifying that it is a pipeline of depth 3, i.e., a pipeline that has three registers between input and output. The next four lines specify the inputs and outputs, apart from a clock and a reset, the unit takes a stream of commands, and produces a stream of computed values. Lines 16–22 define a register containing the current state of the unit. The register is called `state`, is clocked by `clk` and is reset back to the `Mult` state by the `rst` signal. The next state is given by the match expression on Lines 17–22, which retains the current state if the command is not valid, or if it is data to be processed. Otherwise, the state is set to perform addition or multiplication depending on the command.

Lines 24–32 extract the left- and right-hand operands of data commands if present before performing both addition and multiplication. The `trunc` function truncates a variable to the bit-width of the target variable. In the case of the truncations on Lines 31–32, the correct bit-width is inferred by the Spade compiler from the type parameter `T` of the `Stream` result of the pipeline. Line 33 specifies that before doing anything else, the compiler should insert three pipeline registers for all the signals in the design. Using three registers allows the synthesis tool to efficiently map the multiplication into a DSP-block (the dedicated hardware for, among other things, multiplication) in an FPGA. Finally, on Lines 34–37, the output is selected depending on the state, validity of the result, and the sum and product. The pipeline construct ensures that the `state` variable here refers to the state of the unit at the time when the computation was started.

To highlight how the pipelines work, Fig. 7.2 contains an example trace of the unit performing one product, then switching the mode to addition to compute a

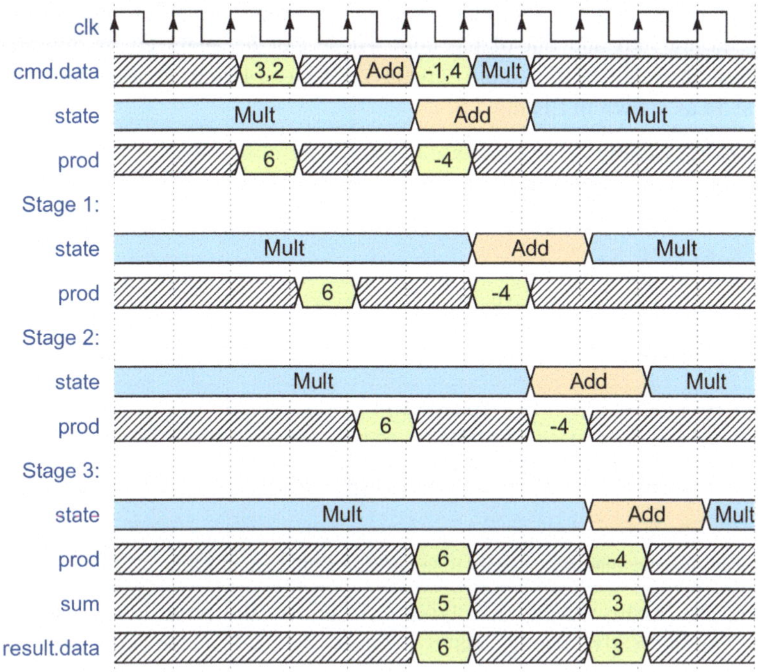

Fig. 7.2 Waveform of the pipeline defined in Listing 7.4 processing a short data stream

single sum before returning to multiplication mode. In the interest of space, the pipelined copies of the `sum` variable have been omitted.

Spade also comes with a build system called Swim.[4] Among other things, this build system runs synthesis and simulation tools, manages dependencies, and supports extensions through plugins. The plugin and dependency system is of particular interest in this chapter, as it allows developers of Spade libraries to bundle WAL analysis passes with their libraries. Additionally, it allows the build system to automatically detect and run these passes when requested.

7.4 Using and Defining Analysis Passes

In this section, we show how the automatic waveform analysis is integrated into a Spade program. We will use the running example from Listing 7.4 and show how it is augmented to both define and use analysis passes.

[4] https://gitlab.com/spade-lang/swim

The primary candidates for analysis in the running example are the two streams of data going into and out of the unit. Those can be analyzed by applying the `wal_trace` attribute to the input stream on Line 14 and to the output stream on Line 34 as shown in Listing 7.5.

```
7  #[wal_traceable(uses_clk, uses_rst)]
8  struct port Stream<T> { valid: bool, data: T } {

12 pipeline(3) main(
13   clk: clock,
14   rst: bool,
15   #[wal_trace(clk=clk, rst=rst)]
16   cmd: Stream<Cmd>
17 ) -> Stream<int<16>> {

34 #[wal_trace(clk=clk, rst=rst)]
35 let result = match state {
36   State::Mult => Stream(out_valid, prod),
37   State::Add => Stream(out_valid, sum)
38 }
```

Listing 7.5 Tracing annotations added to the streams from Listing 7.4

However, because the `Stream` type is defined in the project, not fetched as a library which already includes analysis passes, we need to also define how streams should be analyzed. For most passes, this is done in three steps: annotating the struct to inform the compiler that it is an analyzable struct, writing the WAL code for performing the analysis, and, optionally, writing a Python wrapper for additional Python integration.

7.4.1 Struct Annotation

The first step in enabling automatic analysis is annotating the structs which can be analyzed with the `#[wal_traceable]` attribute as shown on Line 7 of Listing 7.5. This annotation communicates that an analysis pass is available and allows specifying if clock and reset signals need to be provided in addition to the struct signals. With the `wal_trace` and `wal_suffix` signals in place, the Spade compiler will generate `X__valid__proj::main::Stream` and `X__data__proj::main::Stream` signals as well as corresponding signals for the clock and reset signals for each instantiation of the `Stream` struct. For each instance of the stream interface, `X` is replaced by the hierarchical name of the variable defining the stream. The path `proj::main::Stream` is the fully qualified path of the struct in a Spade project and depends on the name of the project and which file the struct is defined in. It is unique and unchanged when a library is added as a dependency to another project.

7.4.2 *WAL Code*

The WAL analysis program for the stream type is defined in Listing 7.6. This program uses WAL's grouping feature together with the Spade integration to automatically run the `stream-utilization` function for every detected stream instance. Lines 1–3 use the `spade-struct` macro to create a list of all groups of signals which are generated by the `wal_trace` attribute. This list is then passed to the `in-spade-struct` macro on Line 5, which performs analysis on each of the stream interfaces. The `in-spade-struct` macro replaces `#field` with the signal corresponding to that field, for example, `#valid` is expanded to `X__valid__proj::main::Stream` where X is the name of the struct instance being analyzed. One metric that is of interest to users of the `stream` interface is the utilization of an instance of this interface, i.e., the percentage of time at which valid data is transmitted. This is measured on Lines 8–14. First, the `whenever` function is used to visit every time step at which the clock rises and the reset is low (Line 8). These are the time steps at which potential data transactions can occur. Next, depending on the value of the `#valid` signal, the `send` or `idle` variables are incremented on Line 9–11. Finally, the fraction of time the interface is sending is calculated on Line 13. This result is then printed together with the name of the interface (stored inside the special variable `CG` which stores the Current Group) (Line 14).

```
1  (define stream-instances
2    (spade-struct proj::main::Stream
3                  [clk rst valid data]))
4
5  (in-spade-structs
6      proj::main::Stream stream-instances
7    (let [(send 0) (idle 0)]
8      (whenever (&& (rising #clk) (= 0 #rst))
9        (if #valid
10         (inc send)
11         (inc idle)))
12     (define time-sending
13       (round (* (/ send (+ send idle)) 100)))
14     (log/analysis CG ": " time-sending "%")))
15
```

Listing 7.6 The WAL implementation of the stream analysis pass in file "utilization.wal"

7.4.3 *Making the Pass Discoverable*

To make an analysis pass discoverable by the swim build tool, it has to be placed inside the Spade project at a predefined location. All analysis passes must be placed inside a *wal* directory in the root of the Spade project.

There are two ways to make an analysis pass discoverable by the swim build tool: (1) the pass can be implemented using plain WAL files, providing a `run-pass` function or macro definition and (2) the pass can be implemented using WAL files and Python wrappers. If one of the two options is implemented, the swim analysis plugin automatically discovers the pass and runs it on every produced wave file.

```
1 class StreamUtilizationPass(WalAnalysisPass):
2   def __init__(self, pass_dir, wavefile):
3     super().__init__(pass_dir, wavefile)
4
5   def run(self):
6     self.wal.eval_str('(eval-file utilization)')
7
```

Listing 7.7 The Python implementation of the stream analysis pass in file "utilization.py"

The analysis pass presented in this example uses the second option of using a Python wrapper class. Therefore, the Spade project contains the two files "utilization.py," which contains the Python wrapper, and "utilization.wal," which contains the WAL program presented in Sect. 7.4.2. The file names of analysis passes are irrelevant for auto-detection by the swim analysis plugin. A Python wrapper must be a class that inherits from `WalAnalysisPass`, which in turn initializes the WAL interpreter, sets up the Spade compiler integration, and provides new WAL functions, for example, for translating Spade values. The constructor for the `WalAnalysisPass` class takes two arguments, the directory of the pass (i.e., the path to the current Spade project) and the waveform that will be analyzed by this pass. WAL analysis passes without a Python wrapper are wrapped automatically, so that they have access to the same Spade integration.

Listing 7.7 shows the Python implementation of a WAL pass that uses the Python wrapper class. Lines 2–3 contain the constructor of the class, which in this case only initializes the base class. The `run` function is defined on Line 5. As most of the logic for this pass is implemented in the WAL program, all the `run` function does is evaluating the *utilization.wal* file on Line 6 which evaluates all the expressions it contains. This is done by evaluating the `(eval-file utilization)` expression using the `eval_str` function of the WAL object, which was already created by the base class.

To run all analysis passes, the swim build tool is invoked as `swim plugin analysis`, which in this case produces the output shown in Listing 7.8 when run on a stream of 1000 random commands with a four in five chance of generating a valid command. As one might expect, the percentage of valid commands is around 80%, while the output, having to switch between modes on roughly one in three valid commands, contains valid data at around 26% clock cycles. This gives a designer a valuable insight: reordering computations to avoid mode switches would improve the throughput of the system.

```
[INFO] Running Stream Utilization Pass on tb.vcd
[ANALYSIS] proj::main::main.cmd_n1: 80%
[ANALYSIS] proj::main::main.result: 26%

```

Listing 7.8 Example output of the analysis

Finally, Listing 7.9 defines an analysis pass without a Python wrapper. The pass prints the operands and results of all valid multiplications, along with the time at which they were performed. The first two lines define the pass and specify the scope of the following signals to avoid having to type out the name of the top module several times. Lines 3–8 filter out all time stamps under which the `prod` variable has valid data. Line 7 calls the `spade/translate` function to query the compiler for the Spade representation of the current value of the `state` variable. Listing 7.10 shows the result of this pass on a short sequence of inputs.

```
1  (defun run-pass []
2    (in-scope 'proj::main::main
3      (whenever
4          (&& (rising~clk)
5              (=~rst 0)
6              (= "Mult"
7                 (spade/translate "state"~state))
8            ~out_valid)
9        (log INDEX ":" ~l "*" ~r "=" ~prod))))
10
```

Listing 7.9 A simple WAL analysis pass, which prints valid values of the `product`variable in Listing 7.4

```
[INFO] 2 : 10241 * 38434 = 58914
[INFO] 6 : 16111 * 1527 = 25497
[INFO] 20 : 63773 * 46328 = 47128
[INFO] 24 : 56822 * 10195 = 27586
```

Listing 7.10 Sample output of the pass defined in Listing 7.9

7.5 Finite State Machine Analysis Pass

The analysis plugin contains a set of predefined analysis passes, one of them providing FSM tracing. This pass analyzes the state distribution and state transitions of state machines. All users have to do to analyze a state machine is to add an `#[fsm]` annotation to the signal containing the state, as shown in Listing 7.11. The FSM analysis pass produces two outputs for each annotated state machine.

```
#[fsm]
reg(clk) state reset(rst: State::Mult()) =
    match cmd { ... }
```

Listing 7.11 Annotating the FSM from the running example to opt into FSM analysis

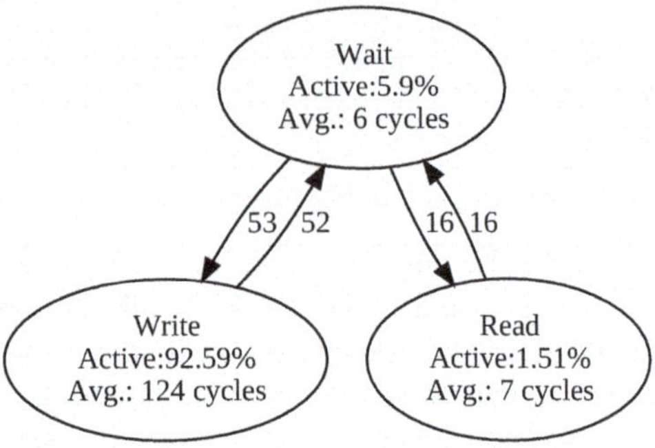

Fig. 7.3 State distribution and state transitions

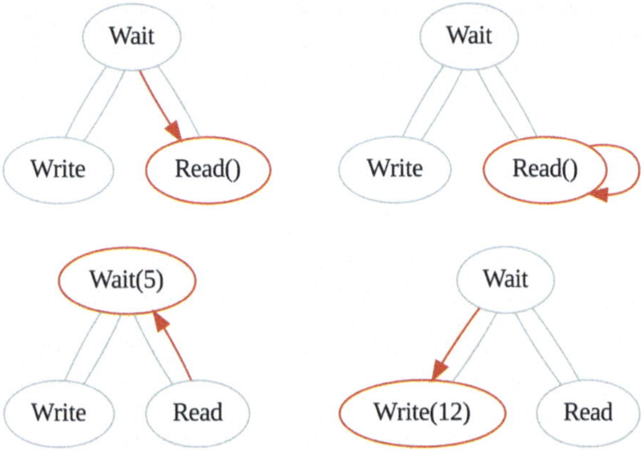

Fig. 7.4 Frames from the generated FSM state transition animation

First, it produces a Control Flow Graph (CFG) like the graph shown in Fig. 7.3. This CFG shows every visited state together with further analysis results about this state. These analysis results include the percentage of time the FSM was in a given state, the average time the FSM spends in a state, and the number of times each state transition was taken. For example, Fig. 7.3 shows the CFG of *wb1* from the previously presented Wishbone masters that were analyzed on the same simulation results as the analysis passes in Sect. 7.2. Of the three states, the *Write* state was active the longest time, and most transitions happened between this and the *Wait* state. Additionally, the 16 transitions between the *Wait* and *Read* states reflect the 16 reads we observed in Sect. 7.2.

Secondly, all state transitions are rendered into sequentially numbered images. A selection of these images is shown in Fig. 7.4 (read from left to right and from top to bottom). These images contain the full control flow graph, in which the current state and the transition that led to this state are colored in red. Additionally, the current state register value is translated using the Spade compiler integration into a string representation of the state name, including its name and current values. This state transition animation allows stepping through each transition, closely following the execution of the FSM.

The FSM analysis is an example of an analysis pass that uses external software packages. Since WAL is fully interoperable with Python, analysis pass authors can use the extensive Python ecosystem to report analysis results in a fitting format or to integrate the analysis into other existing workflows.

7.6 Implementation

The implementation of the WAL integration into Spade is primarily based on the generation of groups of signals with a known suffix, which WAL analysis passes look for. The use of these suffixes allows the WAL passes to be written without much involvement of the compiler. In particular, the suffixes communicate clearly which signals should be traced and which should not and expose individual fields of structs without requiring knowledge of the struct packing. The compiler simply emits Verilog signals with these known suffixes for signals, or parts of signals that should be traced. The compiler exposes a Python API for cases where WAL needs information from the compiler, such as to translate a value from its bit representation to a human-readable Spade value.

The Spade compiler is a multistage compiler with an architecture as shown in Fig. 7.5. The compilation process starts with lexing and parsing to generate an Abstract Syntax Tree (AST). This AST is lowered into a High-level Intermediate Representation (HIR), a process which retains the tree structure of the AST but resolves names and scoping rules, and performs initial semantic analysis. On the HIR, type checking and some transformation passes are performed, and the HIR along with the type information is used to generate a Mid-level Intermediate Representation (MIR). In this step, more semantic analysis is performed, and the tree structure is flattened to a list of simple statements.

Attributes are part of the AST and are baked into the HIR nodes during AST lowering. The WAL-related attributes are type checked and then lowered into dedicated MIR statements during HIR lowering. Finally, those statements are lowered to standard MIR statements, which alias the required signals or sub-signals.

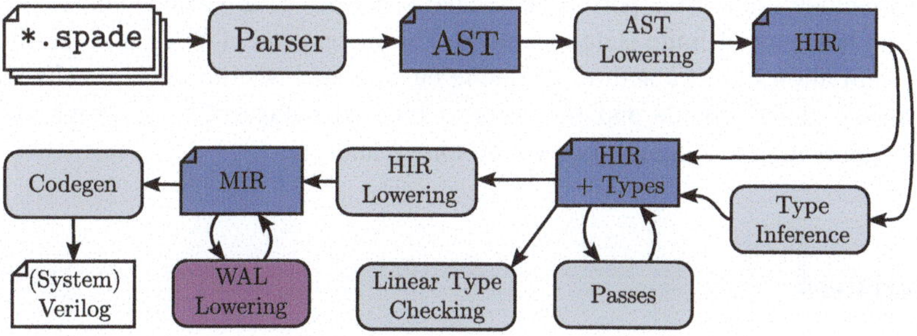

Fig. 7.5 Architecture of the Spade compiler

This has to be done so late in the process, since the final names and types of variables and instances are not decided until the MIR has been generated.

```
let result = match state {
    ...
};
let result__valid__stream = result.valid;
let result__data__stream = result.data;
let result__clk__stream = result.clk;
let result__rst__stream = result.rst;
```

Listing 7.12 Tracing signals added to the streams from Listing 7.4

It is also worth discussing the benefits of integrating WAL with Spade compared to directly using WAL. WAL's primary method of discovering all signals for analysis is, as discussed, via signals with dedicated suffixes. Without integration, these signals must be defined manually. For example, the tracing of the `result` struct in our running example would look like Listing 7.12, as compared to the last block of Listing 7.5. Of course, the same signals would also need to be defined for the `cmd` bus. In addition to being much more tedious to write, this also requires the user to update all the extra signals whenever the fields of a struct are updated. This is extra problematic as the struct being traced in this case is defined in an external library. In the proposed implementation, the compiler also emits an error when the user attempts to trace a struct which is not marked as `wal_traceable`. This ensures that only structs with associated analysis passes are traced, another guarantee which is lost without integration. Finally, without integration with the build tool, it would not be as easy to bundle WAL passes with libraries in a structured way and have those analysis passes run automatically.

7.7 Summary

In this chapter, we showcased how WAL can be embedded inside a HDL. By embedding the WAL waveform analysis language inside the Spade HDL, developers can easily get more information out of their simulation wave forms. A few examples of this include finding the performance characteristics of a wishbone bus, and the run time behavior of finite state machines. The integration of WAL into the Spade ecosystem allows library authors to define custom analysis programs for the types and designs they provide and for users to take advantage of these programs by simply annotating the signals they would like to analyze.

References

1. Amaranth contributors. Amaranth HDL. https://github.com/amaranth-lang/amaranth (2022)
2. C.P. Baaij, Digital circuits in CλaSH. PhD. Thesis, University of Twente (2015)

3. J. Bachrach, H. Vo, B.C. Richards, Y. Lee, A. Waterman, R. Avizienis, J. Wawrzynek, K. Asanovic, Chisel: constructing hardware in a Scala embedded language, in *Design Automation Conference* (2012), pp. 1212–1221
4. A. Izraelevitz, J. Koenig, P. Li, R. Lin, A. Wang, A. Magyar, D. Kim, C. Schmidt, C. Markley, J. Lawson, J. Bachrach, Reusability is firrtl ground: Hardware construction languages, compiler frameworks, and transformations, in *International Conference on Computer-Aided Design* (2017), pp. 209–216
5. J. Kemmerer, PipelineC (2022)
6. L. Klemmer, D. Große, Waveform-based performance analysis of RISC-V processors: late breaking results, in *Design Automation Conference* (2022), pp. 1404–1405
7. L. Klemmer, E. Jentzsch, D. Große, Programmable analysis of RISC-V processor simulations using WAL, in *Design and Verification Conference and Exhibition Europe* (2022)
8. S. Lefebvre, Silice. https://github.com/sylefeb/Silice (2022)
9. P.S. Li, A.M. Izraelevitz, J. Bachrach, Specification for the firrtl language. Technical Report UCB/EECS-2016-9, EECS Department, University of California, Berkeley (2016)
10. OpenCores. Wishbone B4. Technical report (2010)
11. D. Shah, E. Hung, C. Wolf, S. Bazanski, D. Gisselquist, M. Milanovic, Yosys+nextpnr: an open source framework from Verilog to bitstream for commercial FPGAs (2019), pp. 1–4
12. F. Skarman, O. Gustafsson, Spade: an expression-based HDL with pipelines, in *3rd Workshop on Open-Source Design Automation* (2023)
13. F. Skarman, L. Klemmer, O. Gustafsson, D. Große, Enhancing compiler-driven HDL design with automatic waveform analysis, in *Forum on Specification and Design Languages* (2023), pp. 1–8
14. F. Skarman, G. Sörnäs, O. Gustafsson, Spade (2023)
15. SpinalHDL contributors. SpinalHDL. https://github.com/SpinalHDL/SpinalHDL (2022)
16. C. Wolf, Yosys open synthesis suite. https://yosyshq.net/yosys/

Chapter 8
An Interactive Debugging Methodology Based on WAL

This chapter introduces an interactive debugging methodology based on the Waveform Analysis Language (WAL). It utilizes virtual signals, which are signals that are injected into the waveform after simulation at analysis time. These virtual signals are backed by WAL code that is evaluated when a virtual signal is read. By injecting these signals into a trace, existing signals can be replaced. This allows a new debugging methodology in which only small parts of a design are reevaluated after a bug fix, thus removing the need for expensive simulation runs.

8.1 Introduction

In today's rapidly moving industry, design speed and fast iterations are key to getting and staying competitive. When it comes to agility and speed, no industry can keep up with the software industry. But software also relies on fast and efficient hardware as its foundation, and thus hardware companies must find ways to fasten their development cycle to keep up with the demand. In reality, however, hardware development is inherently more complex than software development due to a number of reasons [21]: (1) much higher risk compared to software development, (2) a much more complex process (compiling a binary vs. synthesis, place and route, floor-planning, ...), (3) a far greater dependency on lab time (a single person can write revolutionary new software), (4) far slower and fewer iterations during development (simulations running for multiple days vs. hot-reloading of code in modern software frameworks).

Some of these points are just inherent to hardware design and can hardly be improved (high risk), but there are others that can indeed be improved. In this chapter, our focus is on addressing the long delays between changing an RTL design, waiting for the simulation to stop, and verifying the newly introduced changes. We aim to achieve this by reducing the frequency of new simulation runs and

L. Klemmer, D. Große, *Versatile Hardware Analysis Techniques*,
https://doi.org/10.1007/978-3-031-83093-8_8

providing an integrated environment in which **all** steps of the hardware design process can be performed with fast response times and no clean-cut separations between development, simulation, and verification.

In particular, we want to limit the number of times redundant work is done by providing a fine-grained method for updating existing waveforms. We achieve this by injecting new signals into already existing waveforms that were produced by hardware simulators or by formal tools. This way, only the newly injected signal has to be simulated and not the whole design, dramatically reducing simulation runtime and increasing the iteration speed.

We call these signals **virtual signals**, as they are not included in the original design but are added later, yet they look and behave exactly like the original signals. With virtual signals, existing waveforms can be enhanced, making complicated analysis problems significantly easier to express and enabling developers to iteratively approach their design either for fixing bugs or adding extra logic. Our virtual signals are implemented as an extension to WAL (Chap. 5). By using WAL as a basis, virtual signals can be conveniently defined in a fully fledged programming language. They are therefore even more powerful than traditional signals; for example, they can hold complex data structures like hash maps, or they can call library functions.

To demonstrate virtual signals, we present two detailed case studies in which we show how virtual signals can be used to try out bug fixes without performing new simulations after modifying the logic driving an erroneous signal. We leverage the full potential of WAL, by compiling temporal properties—specified as SystemVerilog Assertions (SVA) [1]—to WAL code, thus using them for bug detection and extremely fast verification of our bug fixes, all while still having access to the full simulation data.

The wish for higher productivity and better development tools is shared by many in industry and academia [15]. Recently, a lot of progress has been achieved by taking modern ideas from software development and transferring them into the hardware domain.

This has been done, probably most famously, by the Chisel HDL [6], a modern hardware generator embedded into Scala. It allows lifting the abstraction of HDL design without losing fine-grained control over every wire and register. Chisel is not the only such new HDL; many other HDLs emerged in the last few years, each bringing its own set of ideas to the table. Notable examples for this are CλaSH [5] which is embedded into Haskell from which it draws many inspirations, Spade [16] which provides specialized constructs for pipelining designs and a build tool inspired by Rust's cargo, and Migen/Amaranth [3].

In the context of modern HDLs and RTL code generation, a range of intermediate languages emerged. These include FIRRTL [14], an RTL intermediate representation that allows, among other things, performing device-specific optimizations and RTL generation, and CIRCT [8], another intermediate representation with related tools.

Previous work into hot reloading of RTL simulations has been presented in [18] and [20]. LiveSim is a framework that allows to update running simulations after changes have been made to the RTL code. The goal of LiveSim is to reduce the edit-run-debug delay, which is the time it takes for a change in the RTL code to become visible in the simulation results. LiveSim achieves a very high simulation performance and reacts rapidly on design changes. However, in contrast to our proposed virtual signals, LiveSim is a very large system on its own and requires adaptions to the development process to integrate LiveSim as a new simulator. Further, it is focused on simulating typical RTL languages. In contrast, virtual signals are not limited to synthesizable constructs or the restrictions of RTL languages. Virtual signals can hold arbitrary data structures, and every virtual signal is in itself a full program. Additionally, virtual signals are tightly integrated into WAL allowing them to interact with other WAL programs. For example, in this chapter, we use an SVA-to-WAL [13] compiler to detect bugs and verify that the suggested fixes are correct. All this is possible within the same environment, and it can be easily extended with other WAL programs, for example, to analyze the performance impact of the changes or to generate visualizations of the changed design. Finally, virtual signals seamlessly integrate into the development process and existing toolchains without requiring any modifications. This is because they solely rely on waveforms, which are already generated at various stages of the hardware design process.

This chapter is based on the works published in [12, 13] and is structured as follows: First, the concept of virtual signals is presented in Sect. 8.2 together with examples and an explanation about how they integrate into WAL. Then, in Sect. 8.3, we briefly present how SVA assertions can be compiled to WAL programs. In Sect. 8.4, we present how virtual signals can be used in a novel design-debug-verification workflow. This novel workflow is evaluated in a case study about using virtual signals while debugging a RISC-V processor in Sect. 8.5. Finally, we conclude this chapter with a discussion of virtual signals, their current limitations, and future research directions in Sect. 8.6.

8.2 Virtual Signals

This section introduces the concept of *virtual signals*. In this section, we introduce the general idea of virtual signals, show how virtual signals can be defined, how they integrate with other WAL functions, and how they can be used to build higher abstraction layers that allow a more convenient use.

The core idea behind virtual signals is that they are additional signals which can be injected into the waveform after simulation time. Compared to "real" signals, the value of a virtual signal is not defined by the loaded waveform, but by a WAL expression that is evaluated when the signal is accessed. This means that virtual signals are expressions with the added context of the simulation time. They can be used in other WAL expressions using all methods that are available to regular signals (e.g., relative evaluation, grouping, scoping) and completely behave like other "real" signals that are contained within waveforms. To implement virtual signals, we extended the available open-source implementation of the WAL interpreter.[1]

8.2.1 Defining Virtual Signals

WAL virtual signals can be defined using the `defsig` function. This function expects two arguments; first the name of the signal that is created and secondly, at least one WAL expression that computes the value of the new signal. Please note that the expressions can be any valid WAL expressions and that virtual signals can hold any valid WAL value (e.g., a signal holding a list of all transactions up to the current timestamp). Listing 8.1 shows how some exemplary virtual signals are created. The first example on Line 1 defines a virtual signal whose value is always equal to the value of the "real" signal *counter* incremented by one. This virtual signal definition corresponds to the Verilog concurrent assignment `assign cnt_plus_one = counter + 1;`. Next, a virtual signal that inverts the *reset* signal to generate an active-low reset is shown on Line 2. Finally, a signal that always holds the index at which a write-enable signal (*tb.we*) was high the last time is shown on Line 3. This signal is defined recursively, if the write-enable is currently high; its value is the current index, else its value is the value this signal had at the previous index. This recursive chain repeats until the write-enable signal is high or the start or end of the waveform is reached. Virtual signals are evaluated lazily, and their values are cached, and thus this recursive evaluation does not add significant runtime cost.

```
1 (defsig cnt-plus-one (+ counter 1))
2 (defsig resetn (- 1 reset))
3 (defsig last-we (if tb.we INDEX last-we@-1))
```

Listing 8.1 Defining virtual signals

8.2.2 Integration of Virtual Signals into WAL

In this section, we describe how virtual signals are integrated into WAL using the example of a simple bus interface that will be extended for debugging purposes.

Suppose a design contains a simple bus using a ready-valid handshaking scheme. To give developers a better overview of the bus activity, we are going to extend each such bus interface in the design with a new virtual signal holding a list of all transactions that have been sent so far. For each transaction, we will store an *address*

[1] https://github.com/ics-jku/wal.

and *data* tuple inside an additional virtual signal called *packets*. This signal will hold a growing list of transactions that have happened on this bus until the index at which the signal is evaluated. If the *rst* signal is high, the virtual signal will be set to hold an empty list (Line 4). If the reset is low, it appends the current transaction to its value if the bus is sending (Lines 5–6), else it keeps its old value (Line 7).

The number of such interfaces in a design can vary depending on the design configuration (e.g., a configuration with two vs. a configuration with three devices), and this affects the number of virtual signals we have to inject. Doing this manually quickly becomes tedious, and therefore, virtual signals should be well integrated with the generic programming features of WAL. In general, from the perspective of WAL programs, virtual signals should behave like "real" signals that are loaded from a waveform. The benefit of this is that they can be directly used by all other WAL features which are already present for "real" signals. This means that they should be available via the `SIGNALS` special variable, they should compose with the grouping and scoping features of WAL, and they should follow the WAL timing semantics (i.e., `step` and relative evaluation).

Our virtual signal implementation satisfies these requirements, and hence we can utilize WAL's full generic programming features to automate the virtual signal creation. This allows us to scan the design for all instances of the bus using the `groups` function[2] (Listing 8.2 Line 1) and create new virtual signals at all discovered locations using the `in-groups` function. Inside the body of the `in-groups` function, signals that start with # and virtual signals are expanded to an absolute name, depending on the currently visited group. Therefore, by creating the virtual signals inside the `in-groups` function, they are automatically name-extended and placed at the correct locations, and thus we can automate the task of creating virtual signals. After defining the virtual signals, we go over the complete waveform once for each interface with the `whenever` function. At each timestamp, we check if a new transaction is added (Line 9–11) to the virtual signal and, if this is the case, print the current index and the *packets* signal (Listing 8.3).

```
1  (in-groups (groups clk rst rdy vld addr data)
2    (defsig packets
3      (if #rst
4        '()
5        (if (&& (rising #clk) #rdy #vld)
6            (append #packets@-1 (list #addr #data))
7            #packets@-1)))
8    (print CG ":") ;; print Current Group
9    (whenever (&& (rising tb.clk)
10                 (! (stable #packets)))
11     (printf "%2d: %s\n" INDEX #packets)))
```

Listing 8.2 Extending bus interfaces with virtual signals

2 For a further explanation of the utilized WAL functions, please see [11].

```
tb.c1.:
12: ((27 232))
20: ((27 232) (65 236))
tb.c2.:
4: ((179 48))
12: ((179 48) (185 52))
16: ((179 48) (185 52) (194 54))
```

Listing 8.3 The transactions on two bus interfaces

8.2.3 *Register and Wire Macros*

Using WAL's macro system, users can extend the language to their needs. In this section, we present two macros that provide a convenient way to define virtual signals.

8.2.3.1 Virtual Wires

The first macro, `wire`, allows defining new combinational signals using a `(wire name expr)` style similar to Verilog. Signals in the expression of a `wire` definition always refer to the signal value at the current index. Listing 8.4 shows how a combinational virtual signal to detect if some clearing operation has to be performed can be created using the `wire` macro.

```
(wire clear (&& cmd_valid (= cmd[1:0] 3)))
```

Listing 8.4 Defining a combinational signal using the `wire`macro

8.2.3.2 Virtual Registers

Now we take a look at defining sequential virtual signals. Sequential logic can reference previous signal values. Typically, in RTL code, signals in expressions like have to be interpreted in two ways. Signal names on the left of the assignment reference the next value of the signal while signals on the right reference the current value. Yet, in the context of virtual signals, this changes slightly as virtual signal expressions always describe the value of the signal at this index, not the value of the signal at the next index. When a WAL virtual signal is evaluated, it is already at the "next" index, so when we want to model a sequential signal, we actually want to reference previous signal values.

```
1 (defsig counter
2   (if (rising clk)
3       (if (|| reset (= counter@-1 5))
4           0
5           (+ counter@-1 1))
6       counter@-1))
```

Listing 8.5 Defining a counter using a virtual signal

This can be achieved manually by using the relative evaluation feature as shown, for example, in Listing 8.5, where the previous signal value is accessed using the `@-1` syntax (e.g., Line 3). However, this quickly makes the code more verbose and unreadable. Therefore, we utilize the macro system built into WAL to provide a convenience macro for defining sequential virtual signals in Listing 8.6. This macro provides a higher-level view on sequential logic than, for example, Verilog and does more than just evaluating the expression at the previous timestamp. Most importantly, it provides a concise way to handle synchronous resets, a feature which is implemented in many modern high-level hardware description languages such as [6, 16, 19]. A macro for asynchronous resets is also possible and would look very similar.

```
1  (defmacro reg [name config expression]
2    "call (reg name [clk [rst init-value]] expr)"
3    (define clk (first config))
4    (define rst (first (second config)))
5    (define rst-value (second (second config)))
6    `(defsig ,name
7       (if (rising ,clk)
8           (if ,rst
9               ,rst-value
10              (reval ,expression -1))
11          (reval ,name -1))))
```

Listing 8.6 A macro for defining sequential virtual signals

The macro expects three arguments: the *name* of the new virtual signal; configuration information about the clock and reset domains, including the reset value (*config*); and the *expression* that defines the value of the new virtual signal.

The clock and reset configuration are passed as a list whose first element is the name of the clock signal and the second element is a tuple containing the name of the reset signal and the reset value. These configuration values are extracted from the argument on Lines 3–5. At their core, WAL macros are full functions, and therefore it is possible to perform any computation in them, including preparing data for the returned expression. The expression returned by the macro is shown on Lines 6–11. First, updating the signal value is guarded by the rising edge detection on Line 7. If the clock is not rising, the value of the signal is assigned to the previous value on (Line 11). When the clock signal is rising, a check is performed if the reset signal is high (Line 8), and if it is, the value of the signal is set to the initial value (Line 9). If the reset is low, the *expression* is evaluated at the previous timestamp, and the result is the new value of this signal (Line 10).

Listing 8.7 shows an implementation of the modulo-6 counter from Listing 8.5 using the new `reg` macro. Compared to the definition in Listing 8.5, the counter definition using the `reg` macro is much more concise and only contains the "core" computation of the counter value. All reset and time handling operations are hidden by the macro.

```
(reg counter [clk [rst 0]]
  (if (= counter 5)
      0
      (+ counter 1)))
```

Listing 8.7 Defining a sequential signal using the `reg`macro

8.3 Compiling SystemVerilog Assertions to WAL

SVA is an industry standard for specifying properties that describe the correct behavior of a system, which is mostly used in formal verification settings. Even though SVA is mostly used in formal verification settings, it is also used during simulation. Checking SVA properties on waveforms was first presented by SAWD in [2]. However, SAWD was not made available, and it is a single-use-case application that cannot be easily integrated into other workflows.

In this section, we propose to compile properties specified in SVA to WAL. This section is based on [13]. We present WSVA, a first implementation of such a compiler.[3] WSVA can compile SVA properties to WAL programs, leveraging WAL's waveform analysis features and its ability to easily act as a back end for other languages. Compiling verification properties is a challenging task and has been studied before, for example, in synthesizing SVA or Property Specification Language (PSL) properties to hardware [4, 7, 9]. By using WAL as a back end, SVA properties can exploit all features available in WAL which often have similar counterparts in SVA. Further, since the compilation target is WAL, the compiled properties are fully integrated into the WAL ecosystem. This means that they can be used by any regular WAL program or in the novel debugging flow presented in this chapter.

Currently, WSVA is a proof-of-concept and in an experimental stage. Therefore, it is not intended to be used for proper verification of a design. However, even in its current form, WSVA allows some interesting use cases: First, it gives people without access to costly commercial tools the possibility to get to know and try out SVA. Second, because it operates on waveforms, it allows to quickly develop properties without having to simulate the same design again. And last, because of its current experimental nature, WSVA can serve as a test bed for new ideas in open-

[3] https://gitlab.com/lklemmer/wsva.

source EDA tools. For example, we plan to integrate WSVA into waveform viewers such as Surfer [17] to provide users a very direct and rich debug and verification environment.

In the rest of this section, we provide a small example of how SVA properties are compiled into WAL programs. Consider the SVA property shown in Listing 8.8. This property checks that on every rising clock edge, an asserted *start* signal results in the counter *cnt* being set correctly to the value of the *value* signal in the following clock cycle.

```
    set_cnt: assert property(
      @(posedge clk)
      disable iff(rst)
      start |=> (cnt == value));
```

Listing 8.8 Simple SVA property example

This SVA property is compiled to the WAL function shown in Listing 8.9. This function checks if the property holds at the current timestamp. Further, another function is compiled which checks if the property holds on the whole waveform. This function is much simpler, though, and we omit it for brevity.

```
1  (defun sva_set_cnt_single []
2    (define was-disabled? #f)
3    (define started-at TS)
4    (define property-status #f)
5    (defun check-disable [] (set! was-disabled? reset))
6    (timeframe
7      (set (property-status
8            (do (check-disable)
9                (cond [start (do (step 1)
10                                (check-disable)
11                                (= cnt value))]
12                      [#t #t]))))
13     (unless (|| property-status was-disabled?)
14             (list started-at TS))))
```

Listing 8.9 WAL code generated for simple property

The variables defined on Line 2–4 track if the disable condition was true some time, when the property started being evaluated, and the result of the property, i.e., if it is violated or if it holds. Next, the `check-disable` function, which is used inside the property to determine if the disable condition is true, is defined. Finally, the generated code of the property is on Line 8–12. First, the code checks if the disable condition is true on Line 8. The implication is compiled into a WAL `cond` expression, shown between Lines 9 and 11. If the antecedent is true (i.e., `start` is high), the time is advanced to the next clock edge (using `step 1`), the disable condition is checked, and the result of the consequent is returned as the result of the implication. If the antecedent is false, the result of the implication is true (we plan to support vacuous results in the future). Finally, if the result was not true and the disable condition was not activated, a list containing the start and end matching times of the property is returned. If the property was true, or if it was disabled, the function returns a null value. This is a simple example. However, it shows how WAL code for complex properties can be constructed by following a set of rules for translating SVA operators to WAL.

In the rest of this chapter, we will use the WSVA compiler to compile SVA properties to WAL which we will use in our proposed design-debug-verification workflow.

8.4 Debugging with Virtual Signals

In this section, we introduce our design-debug-verification workflow based on virtual signals using a simplified but not far-fetched case study. Suppose one of the components of a new system contains a counter that, according to the specification, should wrap around after the value 5. However, during development, a member of the design team misread the specification and implemented the wrap around after the value 4, the nearest power of two. These types of errors are commonly referred to as *off-by-one* errors, and they can occur with remarkable ease. Since bugs of this kind are very common, they are one of the classic bugs in both software development and hardware design.

In the following case study, we will present our methodology based on virtual signals in three steps: (1) finding the bug in the waveform using an SVA property, (2) injecting a fix for the bug into the waveform, and (3) verify the bug fix using the same SVA property.

To find the bug inside the waveform, we compile the SVA property that uncovered it originally to a WAL function. We then inject a new virtual signal into the waveform that fixes the previously uncovered bug. Next, we simulate this, **and only this** signal, and check that it did not invalidate the waveform data by deviating from the original signal before the time at which the SVA property first fails. Finally, we run the same SVA property again to verify that our fix indeed is correct. Please note that all of this happens in the same debugging environment and using one unified language, WAL. The verification engineer does not have to use multiple tools or even simulate the whole design again multiple times during debugging. Only after the engineer came up with a fix that passes the SVA property on the waveform, a new full simulation or other kind of regression run is needed. This way, **time-consuming and workflow breaking simulation is reduced to a minimum**. The WAL shell session containing all steps of this case study is shown in Listing 8.11.

8.4.1 Step 1: Finding the Bug

We start the bug fixing process with the knowledge that a bug was uncovered by one of the SVA properties that have been checked during regression testing. The

```
check_overflow: assert property(
  @(posedge clk)
  disable iff (rst)
  counter == 0 |-> textdollarpast(counter) == 5;
);
```

Listing 8.10 The SVA property that detected the bug

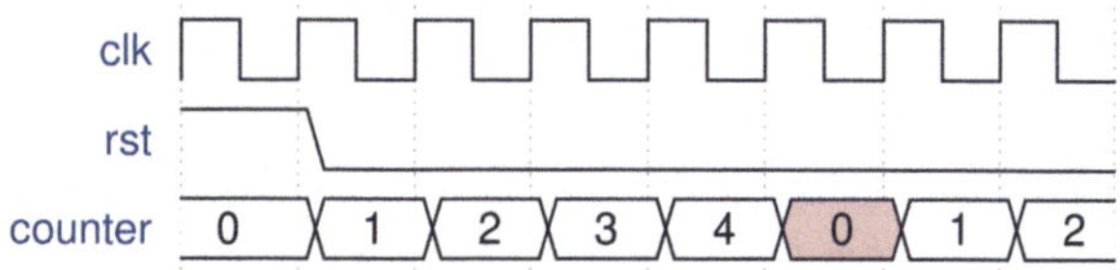

Fig. 8.1 Waveform containing the wrong counter value (red)

first step in fixing a bug using virtual signals is locating the time when the bug first appears in the waveform. For this, we want to utilize the same SVA property that already uncovered the bug. Listing 8.10 shows the SVA property that failed for our design during simulation. The assertion checks if the counter value correctly overflows after it reached the value 5. However, this is not the case at some point in the waveform generated by the simulator.

As WAL is a fully fledged programming language, it is possible to translate an SVA assertion into WAL a program that checks if the assertion holds on a given waveform. We translated the SVA property into a WAL program using an SVA to WAL compiler. This program generates a WAL file containing a function definition *fp* for each property *p* in the SVA file. With this generated WAL file, checking if property *p* holds on the currently loaded task is as simple as calling the function *fp* by evaluating `(fp)`. To use the generated WAL functions, we first have to import them into the running WAL session. In WAL, this is done by evaluating `(eval-file assertions)`, assuming the assertion file was translated into *assertions.wal*. After loading the waveform (Line 1 in Listing 8.11) and the file containing the assertions (Line 2), we can call the *check_overflow* function (Line 3; representing the SVA property check_overflow) to get a list of all timestamps at which the property fails (this list contains tuples containing the time of the first and last matches of the failing assertion). A waveform containing the wrong *counter* value (highlighted in red) is shown in Fig. 8.1.

8.4.2 *Step 2: Injecting the Bugfix*

Having spotted the first timestamp at which the assertion fails, we can analyze the bug. In this case, evaluating the assertion function returns a non-empty list, signaling that the assertion failed at some point. Every entry in the list is a tuple describing the failing assertion. The first element of the tuple is the timestamp at which the assertion first matched, and the second element is the timestamp at which

```
1  >-> (load "counter.vcd")
2  >-> (eval-file assertions)
3  >-> (check_overflow)
4  ((110 110))
5  >-> (whenever (&& (rising clk) (< TS 140))
6          (printf "%3d: %d\n" TS counter))
7    10: 0
8    30: 1
9    50: 2
10   70: 3
11   90: 4
12  110: 0
13  130: 1
14 >-> (reg counter/new [clk (rst 0)]
15       (if (= counter/new 5)
16           0
17           (+ counter/new 1)))
18 >-> (whenever (&& (rising clk) (< TS 140))
19         (printf "%3d: %d %d\n" TS
20                                   counter
21                                   counter/new))
22  10: 0 0
23  30: 1 1
24  50: 2 2
25  70: 3 3
26  90: 4 4
27 110: 0 5
28 130: 1 0
29 >-> (check-integrity counter counter/new 110)
30 >-> (alias counter 'counter/new)
31 >-> (check_overflow)
32 ()
```

Listing 8.11 Injecting a new counter signal into the waveform. ">->" is the WAL shell prompt and TSthe current timestamp

the assertion last matched the trace. Since the *check_overflow* assertion checks only one timestamp, the values of the returned tuple are the equal. In this example, the assertion returns a list with one element pointing to a failure at timestamp 110 (Line 4 in Listing 8.11).

Next, we print all values of the *counter* signal to get an overview of the context in which the bug occurs (Lines 5–13). Here we can see the wrong *counter* value on Line 11, which is 4 due to the bug in the counter, but should be 5. We will now fix this bug by injecting *counter/new*, a new signal that models the counter value with the correct overflow value. For this, we use the previously created *reg* macro that creates a new sequential signal which is sampled on each rising clock edge (Lines 14–17). By creating a new signal *counter/new* with a different name than *counter*, we still have access to the old counter value from the waveform. At this point, we can print the values of both the old and new signals side by side to gain

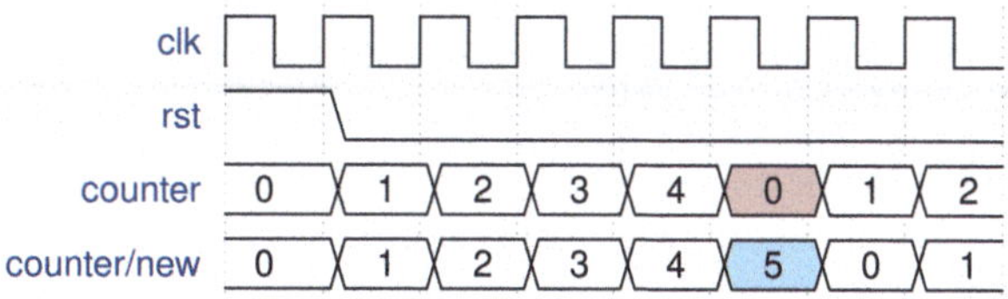

Fig. 8.2 Waveform containing the wrong counter value (red) and the corrected value of the virtual signal (blue)

a deeper understanding of the behavior exhibited by the new signal (Lines 18–28). The new virtual signal is also shown in Fig. 8.2 with the correct maximum value of the counter highlighted in blue.

8.4.3 Step 3: Verifying the Bugfix

At a first glance, the new implementation seems to fix the bug. However, now we need to make sure that our bugfix passes the SVA assertion and that it did not corrupt the other signal values. In general, a corruption can occur if the value of the virtual signal is different from the value of the original signal at any time except the place where the bug occurs. This check, shown on Line 29, is essential, since a mismatch between both signals can invalidate the values of the other signals. This is the case, since we only simulate the new virtual signal; therefore, signals that depend on this value are not updated. In cases in which such a corruption occurs, the simulation data is not valid, and we can make no statement about the correctness of our bugfix. In our counterexample, this integrity check is not needed, since the counter signal is a top-level output of the counter module and no other signal depends on it. However, to show the general principle, we include the check in this example.

Finally, we can check the correctness of the virtual signal using the SVA assertion. Since the SVA assertion checks the correctness of the *counter* signal, we first have to overlay the *counter* signal by the new virtual signal *counter/new*. This is done by introducing an alias for the virtual signal (Line 30 in Listing 8.11). Aliases always have precedence over real or virtual signals; therefore, evaluating *counter* from now on returns the value of *counter/new*. With the old signal overlaid, we can run the SVA property again to see if the new signal corrects the behavior of the old signal (Line 31) on the complete waveform. This time, the assertion returns an empty list, meaning that it did not fail at any timestamp.

8.5 Case Study: RISC-V Hazard Unit

In this section, we repair faulty forwarding logic inside the hazard unit of a pipelined RISC-V processor. Due to a mistake that happened during the wiring of the hazard unit module, the *forwardae* signal, which is a select signal on a multiplexer driving

```
check_forwarding: assert property(
  @(posedge clk)
  disable iff (reset)
  (rs1e == $past(rde)) && regwritem && (rs1e != 0)
    |-> (srca == aluresultm)
);
```

Listing 8.12 The SVA property that detects the wrong forwarding signal value

one of the ALU inputs, *srca*, is undefined in one specific case. If the bug is triggered, the ALU operates on an outdated value and thus can produce the wrong result.

During the verification of the processor, the bug was detected using the SVA property shown in Listing 8.12. This property checks that the *srca* ALU input is the same as the previous ALU result if the register number *rs1* of ALU operand *a* of the instruction currently in the *execute* stage is the same as the result register *rd* of the previous instruction and the previous instruction produces a result that is written to a register other than *x0*. Like in Sect. 8.4, we translated the SVA property to WAL using a SVA to WAL compiler [13].

Listing 8.13 documents a shell session in which the faulty forwarding logic is debugged and fixed. First, the waveform with the bug (Line 1) and the compiled SVA properties are loaded into WAL (Line 2). Next, the SVA property *check_forwarding* is checked against the waveform (Line 3), and a list of the locations at which the property fails is returned (Line 4). Using the *step-to-ts* function, we step forward to the time at which the property fails (Line 5). Here, we check the value of the *forwardae* control signal and see that it is 0 (Line 7). However, since the instructions that are currently in the pipeline produce a hazard, this control signal should be set to 2.

```
1  >-> (load "bug.vcd" t)
2  >-> (eval-file assertions)
3  >-> (check_forwarding)
4  ((90000000 90000000))
5  >-> (step-to-ts 90000000)
6  >-> forwardae
7  0
8  >-> (wire forwardae/new
9         (cond
10          [(&& (= rs1e rdm) regwritem rs1e) 2]
11          [(&& (= rs1e rdm) regwritew rs1e) 1]
12          [else 0]))
13 >-> (alias forwardae 'forwardae/new)
14 >-> forwardae
15 2
16 >-> (check_forwarding)
17 ((90000000 90000000))
18 >-> (wire srca/new
19        (case forwardae
20          [0 rd1e]
```

```
21 |          [1 resultw]
22 |          [2 aluresultm]))
23 | >-> (alias srca 'srca/new)
24 | >-> (check_forwarding)
25 | ()
```

Listing 8.13 WAL shell session documenting the analysis and repair of the forwarding logic

On Lines 8–12, the new logic for the *forwardae/new* signal is defined using the `wire` macro. This logic can detect two kinds of hazards (i.e., possible conflicts with the two previous instructions) and sets the *forwarde* signal accordingly. Now, the original *forwardae* signal is overlaid with the *forwardae/new* signal (Line 13). We can now observe the correct value of *forwardae* (Line 15). Nonetheless, the SVA property still fails at the same timestamp (Line 17). This is because the SVA property checks not the *forwardae* but the *srca* signal. The *srca* signal is driven by a simple 3-Mux that uses *forwardae* as the select signal. Hence, we can easily inject a new virtual signal to propagate the bugfix to *srca* (Lines 18–23). Finally, the SVA property is checked again, this time without failing (Line 25).

8.6 Summary

In this chapter, we take the first step in breaking up the strict separation between design time, simulation time, and verification time. For this, we presented virtual signals which allow injecting new logic into simulation waveforms after the simulation is finished. By utilizing virtual signals, a novel and highly interactive debugging approach emerges, enabling new possibilities for investigation and analysis. Instead of the need to exit the development environment for every minor code change, virtual signals enable the immediate observation and verification of the changes' impact. Likewise, the reverse scenario holds true: In the event of a bug discovered within a waveform, virtual signals facilitate the analysis and resolution of the bug within the same environment.

However, presently, our virtual signal workflow only works under some assumptions. For example, if a signal is overlaid with a virtual signal, in most cases, the virtual signal must behave exactly the same as the original signal for all timestamps up to where the SVA property fails. Additionally, after the values of the two signals diverge, the other values of signals that depend on the changed signal become invalid, as they too have to be now simulated again. We plan to address this in the future by analyzing the cone of influence of a signal. Then, signals that are affected by a virtual signal can be converted into additional virtual signals. Ideally, these steps would have to be performed automatically, for example, by a synthesis step that compiles signals of the original design into virtual signals.

Another direction of future work concerns how virtual signals are defined. WAL is specifically designed to make expressing hardware problems as easy as possible. However, the S-expression and prefix notation make it hard for people not familiar with Lisp languages to pick up WAL. On the other hand, the same S-expression notation also significantly simplifies compiling other languages to WAL.

Alternative front ends for WAL are already available: Including WAWK, which is a waveform analysis language inspired by AWK, and WSVA, the SystemVerilog Assertions front end we used in Sects. 8.4 and 8.5. Both AWK and SVA are already widely used by hardware engineers, providing them an easy entry into the WAL system. Adding a new front end that translates Verilog or VHDL to WAL would make virtual signals available to hardware engineers without any knowledge about the underlying technology. This also highlights a major strength of using WAL as the back end for virtual signals: All projects that utilize WAL are automatically compatible with each other. Further, as WAL is centered around waveforms, which are produced in large quantities during simulation and formal verification, it is completely supplementary to existing toolchains and workflows. In future work, we will investigate how virtual signals can be made permanent by propagating them back up into the original design.

We believe that improvements in productivity are necessary to effectively meet the ever-increasing demands placed on domain-specific hardware [10]. One approach to achieve this objective is by minimizing simulation overhead and streamlining the development process. Even though virtual signals are still limited in their current form, we are confident that they are a promising new direction toward a modern and more interactive hardware development flow.

References

1. *IEEE Std 1800-2017: IEEE Standard for SystemVerilog–Unified Hardware Design, Specification, and Verification Language* (IEEE, New York)
2. A. Alsawi, Sawd: systemverilog assertions waveform based development tool, in *Design and Verification Conference and Exhibition Europe* (2022)
3. Amaranth contributors, Amaranth HDL. https://github.com/amaranth-lang/amaranth (2022)
4. O. Amin, Y. Ramzy, O. Ibrahem, A. Fouad, K. Mohamed, M. Abdelsalam, System verilog assertions synthesis based compiler, in *2016 17th International Workshop on Microprocessor and SOC Test and Verification* (2016), pp. 65–70
5. C.P. Baaij, *Digital circuits in CλaSH*. PhD. Thesis (University of Twente, Twente, 2015)
6. J. Bachrach, H. Vo, B.C. Richards, Y. Lee, A. Waterman, R. Avizienis, J. Wawrzynek, K. Asanovic, Chisel: constructing hardware in a Scala embedded language, in *Design Automation Conference* (2012), pp. 1212–1221
7. M. Boule, Z. Zilic, Automata-based assertion-checker synthesis of PSL properties. ACM Trans. Des. Autom. Electron. Syst. **13**(1), 1–21 (2008)
8. CIRCT contributors, CIRCT / Circuit IR Compilers and Tools. https://github.com/llvm/circt/tree/main/ (2023)
9. S. Das, R. Mohanty, P. Dasgupta, P.P. Chakrabarti, Synthesis of system verilog assertions, in *Proceedings of the Design Automation & Test in Europe Conference*, vol. 2 (2006), pp. 1–6
10. J.L. Hennessy, D.A. Patterson, A new golden age for computer architecture. Commun. ACM **62**(2), 48–60 (2019)
11. L. Klemmer, D. Große, WAL: a novel waveform analysis language for advanced design understanding and debugging, in *ASP Design Automation Conference* (2022), pp. 358–364

12. L. Klemmer, D. Große, Towards a highly interactive design-debug-verification cycle, in *ASP Design Automation Conference* (2024), pp. 692–697
13. L. Klemmer, D. Große, WSVA: a SystemVerilog Assertion to WAL compiler, in *Workshop on Open-Source Design Automation* (2024)
14. P.S. Li, A.M. Izraelevitz, J. Bachrach, Specification for the firrtl language. Technical Report UCB/EECS-2016-9, EECS Department (University of California, Berkeley, 2016)
15. A. Rautakoura, T. Hämäläinen, Does SOC hardware development become agile by saying so: A literature review and mapping study. ACM Trans. Embed. Comput. Syst. **22**(3), 1–27 (2023)
16. F. Skarman, O. Gustafsson, Spade: an expression-based HDL with pipelines, in *3rd Workshop on Open-Source Design Automation* (2023)
17. F. Skarman, O. Gustafsson, L. Klemmer, Surfer 0.1.0 (2024)
18. H. Skinner, R. Trapani Possignolo, S.-H. Wang, J. Renau, Livesim: a fast hot reload simulator for HDLS, in *IEEE International Symposium on Performance Analysis of Systems and Software* (2020), pp. 126–135
19. SpinalHDL contributors, SpinalHDL. https://github.com/SpinalHDL/SpinalHDL (2022)
20. S.-H. Wang, R.T. Possignolo, H.B. Skinner, J. Renau, Livehd: a productive live hardware development flow. IEEE Micro **40**(4), 67–75 (2020)
21. N. Weste, D. Harris, *CMOS VLSI Design: A Circuits and Systems Perspective* (Addison-Wesley Publishing Company, New York, 2010)

Chapter 9
Netlist Optimization Using Formal Methods Under External Constraints

This chapter presents a netlist optimization methodology based on formal methods. External constraints are constraints on the inputs or outputs of a netlist. Such a constraint can be, for example, the exclusion of specific values for a given input signal. This chapter presents how these external constraints can be efficiently used to reduce the size of the netlist by removing gates that, considering the external constraints, are not having an effect on the output. The presented methodology is evaluated on several benchmarks, including combinational netlists and RISC-V processors.

9.1 Introduction

After a design has passed the verification phase, it is synthesized into a gate-level netlist. In the realm of Very Large-Scale Integration (VLSI) design, synthesis is the process of converting a high-level hardware description, usually provided in the form of a behavioral Register Transfer Level (RTL) design, into a gate-level representation. In the early days, synthesis referred only to the direct transformation into circuitry. Therefore, the term *logic synthesis* was coined [7] to emphasize both compilation and logic optimization of a design description during conversion into a netlist. The focus of this chapter is in particular on the second part, i.e., logic minimization of circuit netlists.

The 1980s saw a significant evolution in logic synthesis algorithms, where the treatment of Don't Cares (DCs) became more advanced and integral to the optimization process (see e.g., [5, 21]). The computation of *Internal Don't Cares*, which arise from the structure of the netlist itself, and their use for minimization has been integrated in all logic synthesis tools. However, this is not the case for *External Don't Cares* [13]. At their core, external DCs are external constraints on inputs that will not appear during runtime. We will provide more details about this in the next

L. Klemmer, D. Große, *Versatile Hardware Analysis Techniques*,
https://doi.org/10.1007/978-3-031-83093-8_9

section. In contrast to internal DCs, external DCs are given by the environment or explicitly by the user. Typical examples for external DCs include restrictions on the values of input operands, instruction subsets used by a processor for specific applications, or limited operation modes of an integrated IP block.

Although there have been attempts to unify the characterization of different DC types [16], targeting external DCs presents a challenge, because the optimization problem changes from considering a completely specified Boolean function to optimizing a Boolean relation. Additionally, it has been reported in [13] that there are currently no open-source synthesis tools available that accept external DCs.

From the practical side, [4] introduced the Property-Driven Automatic Transformation (PDAT) framework which targets the very specific application of eliminating the logic of unneeded instructions in a processor realizing an ISA. PDAT employs PC [3, 20, 23] to detect gates that are guaranteed not to toggle for the reduced ISA. These gates can be removed from the netlist. However, [4] has several shortcomings: Relevant information such as formalization, used properties, and verification directives are not or only partially given, run-time is not reported, only commercial tools are used, and most importantly, PDAT is not available open-source.

In this chapter, we propose an optimization approach that utilizes formal verification methods to reduce the number of gates in a netlist. Our approach extends the previous work by showing how to formulate Property Checking (PC) and Equivalence Checking (EC) [15, 18] problems to determine if a gate is redundant under given external constraints. For this, we formalize the respective formal problems and introduce the rules which have to be checked/applied per gate.

We have implemented our approach as the tool Formal SYNthesis (FSYN). FSYN supports netlist optimization using three different back ends, including two back ends based on PC and one back end based on Equivalence Checking (EC). One of the main goals of FSYN is that it is implemented primarily using open-source tools and that FSYN itself is available open-source.[1] This is especially important considering the highly specialized tools required for the FSYN flow, namely, logic synthesis and formal verification. Until recently, these tools were only available from commercial vendors, which posed a very high entry barrier for smaller companies, researchers, and hobbyists. Just recently, it became feasible to perform all steps of the hardware design flow including synthesis, place and route, and formal verification using only free and open-source tools. This is in large part, thanks to Yosys [22], which acted as a crystallization point for the open-source EDA community. FSYN uses only open-source tools, in particular Yosys [25], ABC [6], WAL (Chap. 5), and Z3 [9]. In the experiments, we demonstrate that FSYN can achieve substantial reductions in the number of gates for different circuits under given external constraints.

This chapter is based on the results published in [10] and [12] and is structured as follows. First, Sect. 9.2 formalizes our optimization approach based on PC and EC

[1] FSYN is available at: https://github.com/ics-jku/fsyn.

and provides illustrative examples. Then, we introduce the high-level FSYN flow, and it's individual phases in Sect. 9.3. In Sect. 9.4, we present FSYNs distributed rule checking back end based on PC, and in Sect. 9.5, we evaluate FSYN on designs ranging from combinational circuits up to full RISC-V processors.

9.2 Netlist Optimization Under External Constraints

In this section, we first give some background on logic synthesis and DCs (Sect. 9.2.1). Thereafter, we present a generic formalization of the proposed optimization approach in terms of PC and EC problems (Sect. 9.2.2). This is supported through the use of simple examples.

9.2.1 Logic Synthesis, Don't Cares, and External Constraints

Toward the end of the 1980s, techniques for optimizing multilevel logic became increasingly prominent. Examples include algebraic manipulation, factoring, common sub-expression elimination, and functional decomposition to minimize logic complexity [11]. As these methods advanced, the use of DCs became pivotal in the process of logic optimization. DCs offer considerable flexibility for optimization algorithms [5, 21]. Typically, *Internal Don't Cares* and External Don't Cares (XDCs) are distinguished. Internal DCs result from the netlist structure in the presence of reconvergent paths. They are divided into Satisfiability Don't Cares (SDCs)and Observability Don't Cares (ODCs). In essence, SDCs occur when certain input combinations are not generated for a node, while ODCs arise when the output value of a node is not important under specific conditions. The computation of both has been formulated using incompletely specified functions also called permissible functions [2, 19]. Significant improvements in terms of scalability became possible by using simulation and Boolean Satisfiability Problem (SAT) (see, e.g., [17]).

As already mentioned in the introduction, XDCs are given by the environment or explicitly by the user. Lee et al. [13] divides XDCs further in two classes by extending the definitions of SDCs and ODCs to the input and output boundaries, respectively. A value assignment to the primary inputs that will never appear (which extends the definition of an SDC to the inputs) is an XSDC. Similarly, external ODCs, termed XODCs, are conditions under which some primary output values are not of interest. In this chapter, we focus on XSDCs. XSDCs can be easily expressed as formal constraints on input signals. From the practical perspective, users can leverage well-known property languages to specify these *external constraints* in form of assumptions. In the simplest case, these assumptions include or exclude values on the primary inputs, but our proposed approach is not limited to this. In

Table 9.1 Gate redundancy rules

Gate	Rules
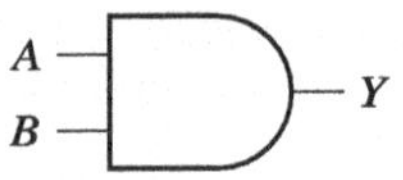	R_AND_Y0: $Y = 0$ R_AND_Y1: $Y = 1$ R_AND_YA: $A \Longrightarrow B$ R_AND_YB: $B \Longrightarrow A$
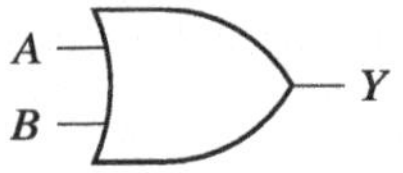	R_OR_Y0: $Y = 0$ R_OR_Y1: $Y = 1$ R_OR_YA: $B \Longrightarrow A$ R_OR_YB: $A \Longrightarrow B$
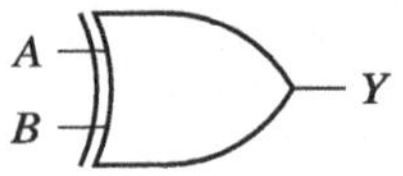	R_XOR_Y0: $Y = 0$ R_XOR_Y1: $Y = 1$ R_XOR_YA: $\overline{B}$ R_XOR_YB: $\overline{A}$
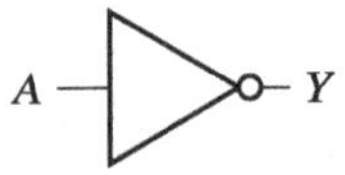	R_NOT_Y0: $Y = 0$ R_NOT_Y1: $Y = 1$

the following, we present the formalization of the proposed optimization approach together with simple examples demonstrating the optimization potential.

9.2.2 Formalization

Given a netlist of a circuit, we are interested in optimizing the netlist under given external constraints. More precisely, our cost function is the number of gates, i.e., we aim to remove gates from the netlist, and thus we are optimizing for area.

First, we introduce some general notations which we use in the remainder of this chapter:

- N denotes a netlist,
- g_i denotes a gate from N, and
- E denotes the user-specified external constraints.

The proposed optimization approach aims to determine if a gate is redundant under specific external constraints. We follow the idea from [4, 12] and capture this gate redundancy formally with ***rules***[2] to be checked and applied to the current gate g_i. Table 9.1 lists the rules for the 2-input gates AND, OR, XOR, and the 1-input NOT gate. In general, a rule in Table 9.1 is defined as R_<gate>_Y<s>: *condition* where (i) the symbols before the colon constitute the name of the rule, and (ii) if the *condition* holds for the gate <gate> (we also say the rule holds), then the output Y

[2] We use the term rules as we look at the problem in a more general way than [4, 12].

is replaced by <S>. Let us consider the AND gate with the output Y and the inputs A and B as an example: For this gate, there are four different rules. For instance, rule R_AND_Y0 describes if $Y = 0$ holds, then the AND gate can be removed and its output Y is replaced by the constant 0. Another example is the rule R_AND_YA which describes if $A \implies B$ holds for the AND gate, then Y is replaced by the input A; in other words, the AND gate can be removed and the output Y is directly connected to input A.

Based on these rules, we formulate the core step of the optimization problem under external constraints, the **gate-redundancy check for gate g_i utilizing PC** as

$$G_{PC}(g_i) = \{E^k \wedge \bigwedge_{j=0}^{k-1} T(s_j, s_{j+1}) \wedge P^k(g_i) \quad | \, P^k(g_i) \in Rules(type(g_i)) \, \} \tag{9.1}$$

where k is the number of time frames ($k = 1$ in the combinational case), E^k are the external constraints, $T(s_j, s_{j+1})$ is the transition relation, and $P^k(g_i)$ is a concrete rule from all possible rules of the gate g_i of $type(g_i)$ in form of a property; please note that exponent k in E^k and $P^k(g_i)$ is used to make clear that the properties have to be considered from 0 to k. Let us look at this formalization: If g_i is, for instance, an AND gate, $G_{PC}(g_i)$ consists of four PC problems. To perform optimization for the complete netlist N, $G_{PC}(g_i)$ must be checked for all gates of N.

We now present a concrete example to demonstrate the concepts.

Example 9.1 We consider a 2-bit Ripple-Carry Adder (RCA) which adds up the binary numbers a_1a_0 and b_1b_0 and produces the result $s_2s_1s_0$ and has no carry in. The truth table for a 2-bit RCA is shown in Table 9.2. Figure 9.1 depicts a classical

Table 9.2 Truth table of 2-bit ripple-carry adder

a_1a_0	b_1b_0	$s_2\ s_1\ s_0$
00	00	000
00	01	001
00	10	010
00	11	011
01	00	001
01	01	010
01	10	011
01	11	100
10	00	010
10	01	011
10	10	100
10	11	101
11	00	011
11	01	100
11	10	101
11	11	110

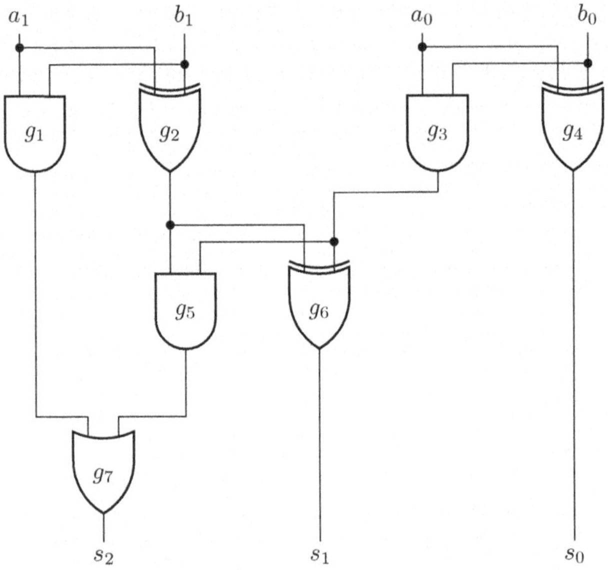

Fig. 9.1 2-bit ripple-carry adder

netlist implementation using a full adder (consisting of the two half adders g_1, g_2 and g_5, g_6 and OR gate g_7) and another half adder (gates g_3, g_4) for the lower bits.

If we now consider an external constraint that operand b is restricted to only even numbers, formally $b_0 = 0$, the circuit can be deeply optimized. The result is shown in Fig. 9.2. Note that in the truth table (Table 9.2), only the light blue lines remain under this external constraint. Let us consider the AND gate g_3 in Fig. 9.1 as an example. For this gate, the rule R_AND_Y0 holds; hence, g_3 is redundant and can be removed; as a result, the output is connected to the 0 constant.

Another example is the XOR gate g_4 with inputs $B = a_0$ and $A = b_0$. Due to the external constraint $b_0 = 0$, which means $\overline{A} = \overline{b_0} = \overline{0} = 1$, rule R_XOR_YB holds and therefore a_0 is directly connected to s_0. The same argumentation holds for gate g_6, and thus the output of g_2 goes directly to s_1. Finally, for the OR gate g_7, the rule R_OR_YB holds and therefore the output of g_1 is connected to s_2. Please note, as we use PC, i.e., formal reasoning, the order in which we check the rules is not relevant for correctness.

Example 9.2 Again, we consider a 2-bit RCA. However, we now have the external constraint that the operand b is restricted to only odd numbers ($b_0 = 1$). In this case, only a single gate (gate g_3) can be replaced by one of its inputs (a_0), since rule R_AND_YB holds for g_3. The result is shown in Fig. 9.3.

Both examples show that the RCA can be optimized under external constraints. As expected, the number of redundant gates depends on the circuit and the concrete external constraints.

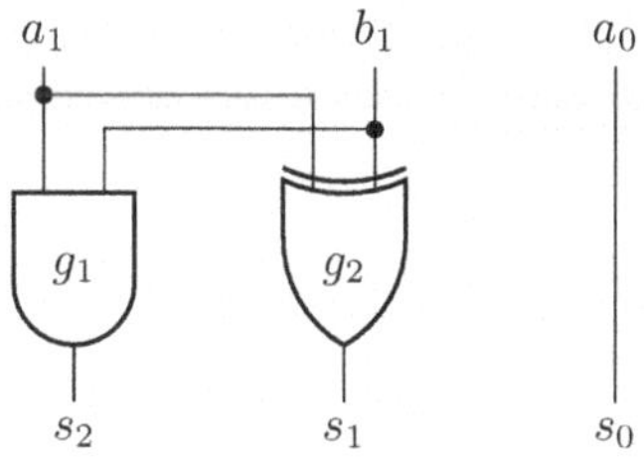

Fig. 9.2 Optimized 2-bit ripple-carry adder under external constraint $b_0 = 0$

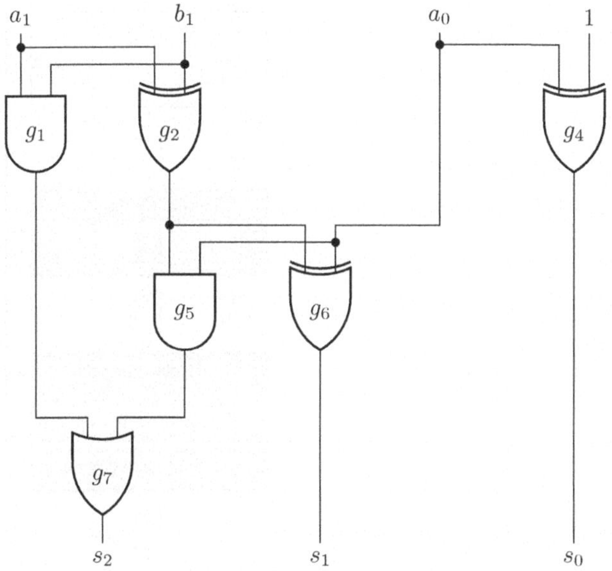

Fig. 9.3 Optimized 2-bit ripple-carry adder under external constraint $b_0 = 1$

We introduce an alternative formulation for the optimization problem at hand. The concept of the presented rules remains the same. However, instead of utilizing PC and describing the rules as properties, we now apply each rule for the gate g_i and create an EC problem.

We can formulate the core step of the optimization problem under external constraints, the **gate-redundancy check for gate** g_i **utilizing EC** as follows:

$$G_{EC}(g_i) = \{E \wedge Miter(N, \hat{N}(g_i)) \mid \hat{N}(g_i) \in Rules(type(g_i))\} \tag{9.2}$$

where E are the external constraints and $Miter(N, \hat{N}(g_i))$ forms a miter. This miter is built for each rule of gate g_i of $type(g_i)$ by applying the rule on the netlist (for instance, by replacing the output of g_i by a constant or connecting the output of g_i to the input A). Recall, if g_i is an AND gate, then four different miters are created. For the other gates, see Table 9.1. Also in this formalization, for optimizing the complete netlist N, all EC problems for each gate have to be solved.

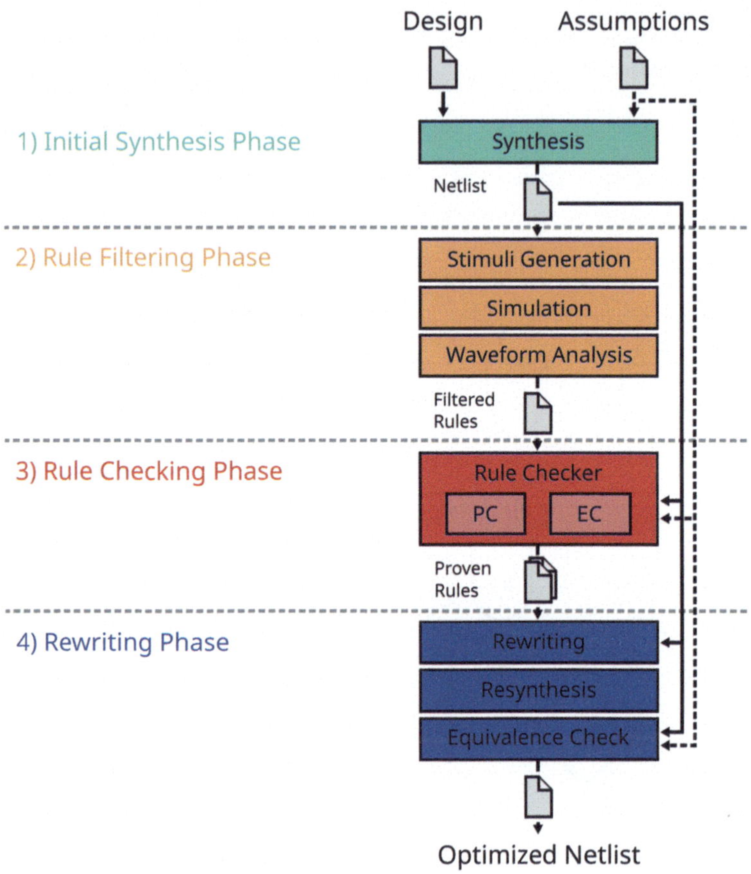

Fig. 9.4 The high-level flow of FSYN

The PC and EC formalizations presented so far allow to optimize a given netlist under external constraints. It is evident that the underlying formal instances can be solved concurrently. In the following section, we introduce our implementation FSYN and discuss several improvements for scalability.

9.3 FSYN Optimization Flow

A high-level flow of FSYN is depicted in Fig. 9.4. To perform optimizations on a design using FSYN, two inputs are required, the *design* that should be optimized and a file containing the input *assumptions* that describe the external constraints, which form the basis for the optimization.

In general, FSYN is split into four distinct phases. In the first *Initial Synthesis Phase* (teal), the design is synthesized into a netlist. This initial synthesis is the first step of the netlist optimization, and it serves as a way to transform designs, which may be in SystemVerilog, VHDL, or in one of the netlist formats supported by Yosys, into a common format that is supported by the rest of the flow.

In the second *Rule Filtering Phase* (orange), FSYN filters the list of all possible optimization rules. As a rule of thumb, there are 4 possible optimizations for every gate in the netlist. Since formal approaches can run into capacity problems with increasing design size, we employ a simulation-based approach to eliminate as many rules as possible by disproving their validity through random simulation. The rule filtering is performed in three steps: First, random input stimuli, which fulfill assumptions, are generated. Then, the stimuli are simulated on the netlist for a certain period of time. Finally, all gates in the resulting waveform are analyzed to find optimization rules that have been refuted. The result of the second phase is a list of filtered optimization rules which have to be formally checked by the formal *Rule Checker*.

The next phase is the *Rule Checking Phase* (red). The previous filtering phase could only disqualify rules based on counterexamples, not prove that a rule is a valid optimization. For this, the rule checking phase employs the formal checks which are the core of FSYN. The rule checker supports three alternative rule checking back ends. The first back end, presented in more detail in Sect. 9.4, is based on PC and utilizes only open-source tools. The second, EC-based back end, uses exclusively open-source tools too. Finally, a PC-based back end using a proprietary commercial PC tool is available as well. Broadly speaking, the PC back ends attached properties to all gates which are then checked. In contrast to this, the EC back end directly performs the optimization first and afterward checks if the resulting netlist is still equivalent to the original netlist under the assumptions. The result of the third phase is a list of proven optimizations.

Finally, in the *Rewriting Phase* (blue), the netlist is rewritten by applying the proven rules to it. After the netlist is rewritten, it is synthesized again to perform some cleanup work (e.g., unnecessary assignment chains, which can be generated by FSYN, are shortened). As a last step, a final equivalence check is performed between the optimized netlist and the synthesized netlist.

The rest of this section presents the four phases in greater detail.

9.3.1 Initial Synthesis Phase

In the initial phase, the design is synthesized using Yosys into a gate-level netlist. This phase serves two main purposes: First, it brings the design to the gate level at which FSYN operates. Since all input designs are going through this initial synthesis phase, FSYN supports a wide variety of input formats including SystemVerilog, VHDL, or other netlist formats supported by Yosys. Second, it unifies the netlist into a known and fixed format, which makes sure that all netlist components are supported by FSYN. The resulting netlist is used by all following steps of the FSYN flow.

9.3.2 Rule Filtering Phase

Applying an optimization rule to a netlist requires proving that the corresponding condition of the rule (cf., Table 9.1) always holds under the external constraints. As explained in Sect. 9.2.2, FSYN uses formal methods to perform this check. However, proving can be much more costly than refuting. For the latter, only a counterexample is needed, which, in the best case, can be easily found by simulation. In the context of this chapter, there is just one challenge, namely, that the stimuli we use for simulation must fulfill the external constraints (given as assumptions). We formulate the problem of finding valid stimuli as an SMT problem over the inputs of the design. This SMT problem consists of the external constraints and an always failing assertion. When solving the problem, we get a counterexample with a set of valid inputs that fulfill the external constraints. To generate multiple inputs, we update the problem to exclude already seen input combinations using the technique described in [8].

We then use the generated stimuli to simulate the netlist to produce waveforms with the behavior of every gate in the design during the simulations. Now, for every gate and every rule, we try to find a timestamp inside the waveform using WAL (Chap. 5) at which the rule is refuted.

An exemplary WAL expression that checks if the constant 0 rule is refuted is shown in the following listing:[3]

```
(> 0 (count (! (= top.s[2] 0))))
```

In the concrete WAL expression, we count how often the signal `top.s[2]` is not 0 on the waveform, in other words, if we see at least a single 1, the rule is refuted. For other rules, the subexpression `(= top.s[2] 0)` would be replaced accordingly, and the outer expression remains unchanged.

In FSYN, the rule filtering phase is optional and can be used to reduce the number of required formal proofs. The gate filtering also employs an SMT solver but compared to the PC and EC problems in the third phase, the stimuli generation scales only with the number of inputs, not with the number of gates. Analyzing the simulation waveform, the second step of the rule filtering phase, scales, in theory, quadratically with the number of gates in the design and the number of stimuli generated. However, generating only a few stimuli is often sufficient to eliminate a large portion of the gates and thus to significantly reduce the rule checking time. We demonstrate the effectiveness of the rule filtering approach in Sect. 9.5.6.

---

[3] `(count expr)` counts and returns at how many timestamps an expression `expr` evaluates to true.

9.3.3 Rule Checking Phase

In this section, we describe the two rule checking implementations of FSYN. FSYNs architecture allows having multiple back ends for the rule checking phase. Currently, three back ends are implemented: two using PC and one using EC.

The PC back ends utilize by a front end for the formal verification capabilities of Yosys [25] and a commercial tool for solving the properties. The property checking in the open-source-based PC back end is split into many smaller tasks. Each task is a subset of rules for which properties will be checked in a single call to sby of Yosys.

The standard configuration is a task size of 80 gates (see analysis in [12]) with at most 4 properties/rules to be checked per gate, i.e., a single PC instance contains up to 320 properties. These tasks can be parallelized by leveraging the available threads of the local machine as well as by distributing the computation to other computers. A detailed description on the two PC back end is given in Sect. 9.4.

The EC back end is currently in an experimental stage. As presented in Sect. 9.2.2, the EC back end rewrites the netlist eagerly and then checks if the rewritten netlist is equivalent to the original netlist using Yosys equivalence checking. The EC back end allows local parallelization. However, it has yet to be integrated into the distributed system used by the open-source PC back end presented in Sect. 9.4.

9.3.4 Rewriting Phase

The rewriting phase is the final phase of FSYN. In this phase, the proven rules are applied to the netlist. The rewritten netlist is then optimized again using Yosys to clean up remnants of the rewriting phase. For example, FSYNs rewriting can leave chains of assignments ($a = b = c$) which can be removed by Yosys. Finally, an equivalence check between the cleaned rewritten netlist and the original synthesized netlist is performed.

9.4 FSYN: Open-Source PC Back End Leveraging Compute-Clusters

The FSYN flow can utilize different back ends to formally check the optimization rules. We implemented three back ends, two using only the previously mentioned open-source tools and one back end using a commercial PC tool. Adding new back ends is straightforward, as each back end is a class that has to provide only three functions: (1) *inject*, which adds the assumptions to the design; (2) *bind*, which binds the properties to the gates; (3) *work*, which performs the PC.

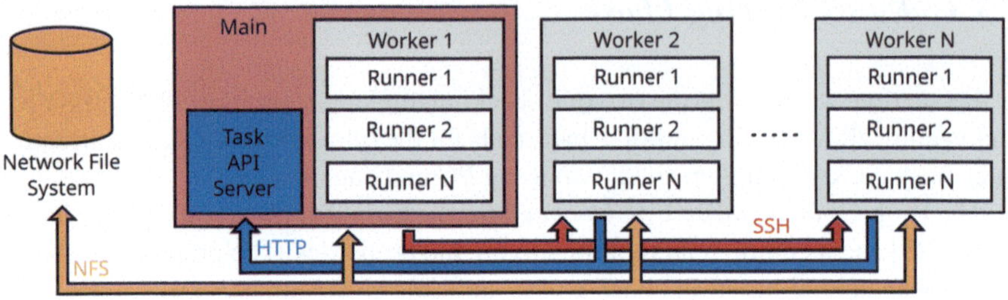

Fig. 9.5 The distributed architecture of FSYN

In the remainder of this section, we will focus only on the open-source PC back end, as this back end is freely accessible and allows our proposed large-scale distributed flow.

Our *open-source PC back end* leverages sby and allows distributing PC onto multiple machines, if the machines are connected in a specific setup. An overview of this setup is shown in Fig. 9.5. The number of machines in the setup is not limited. However, one of the machines must act as the main coordinating process (light red box around Worker 1). This process is the entry point to FSYN and performs the initial synthesis, rewrites and resynthesizes the netlist, and performs the equivalence checking. It also provides the task API, to which the workers connect during PC to get new tasks. To this end, a port on the main machine must be open such that they can make requests to the task API (blue lines).

Further, all machines must have access to a shared directory over a network file system (orange lines and cylinder). This shared directory is required to exchange the netlist, the assumptions, and the redundant gates. Finally, ssh connections between the machines must be possible, so that the coordinating process can start the clients on all workers (red lines).

9.4.1 Configuration and Usage

Each hardware design is unique and distinct from one another. Therefore, FSYN provides configuration features: First, the *config.ini* file is the main place for declaring designs that should be made available to FSYN. To add a new design, a new section has to be added to the config file. Listing 9.1 shows an exemplary configuration for the Ibex processor. A new section is declared by writing the name inside brackets. The filename and the name of the top module are specified using the keys *design* and *top*, respectively. Next, the file holding the assumptions is specified using the *assumptions* key. If the design is sequential, the reset signal and its polarity can be set via the *reset* and *active-low* keys. Additionally, the number of times a sequential design is unrolled and the number of cycles that are skipped by the PC tool are declared using the *unroll* and *skip* keys.

```
[ibex]
design = ibex_top.v
top = ibex_top
assumptions = ibex_assumptions.v
reset = rst_ni
active-low = 1
unroll = 8
skip = 2
```

Listing 9.1 FSYN configuration file

The next configuration file defines which machines are used by FSYN. This is done using a *workers.json* file, which must contain a JSON dictionary with the hostname of the workers being the keys and the number of parallel threads on a worker being the values. An example of this is shown in Listing 9.2. Note that the main machine that runs the coordinating process does not have to be, but can be, included in this list; it only has to be connected to the workers like discussed above.

```
{
    "worker1": I,
    "worker2": J,
    ...
    "workerN": K
}
```

Listing 9.2 FSYN worker configuration file

9.4.2 Assumption Specification

Before FSYN can be applied to a design, users have to provide a file containing the assumptions they would like to make about the input data. An exemplary assumption file is shown in Listing 9.3. In this example, we want to remove some instructions from a RISC-V processor. To accomplish this, we write an assumption about the *signal* coming into the core from the instruction memory. In this assumption, we enable specific instructions that should be included in the final netlist by only allowing certain legal values for the *op*, *funct3*, and *funct7* values of RISC-V instruction words. Now, removing an instruction is as simple as commenting out a line from the assumption. For example, the assumptions below enable all RV32I instructions[4] except the *slli* instruction (Lines 6–8).

```
1 always @* begin
2   assume ((instr[6:2] == 5'b01101) // lui
3          || (instr[6:2] == 5'b00101) // auipc
4          || (instr[6:2] == 5'b11011) // jal
5          ...
```

[4] We omit most assumptions for brevity.

```
        // || ((instr[31:25] == 7'b0000000)
        //      && (instr[14:12] == 3'b001)
        //      && (instr[6:2] == 5'b00100)) // slli
        || ((instr[31:25] == 7'b0000000)
             && (instr[14:12] == 3'b101)
             && (instr[6:2] == 5'b00100)) // srli
        ...
        );
  end
```

Listing 9.3 An exemplary assumption file for removing instructions from a RISC-V core

9.4.3 *Algorithm*

In this section, we present the main algorithm of FSYNs PC and open-source-based back end. The algorithm consists of three steps which are preparation, distributed execution, and finalization. The first step, which is performed before PC, consists of synthesizing the design into a netlist and dividing the PC problems into individual tasks that can be distributed over multiple machines. Then, in the second step, the tasks are distributed to define worker machines, and the PC problems are generated by injecting the rules into the netlist and solved to check which optimization rules apply. Finally, in the third step, the netlist is rewritten based on the PC results, the netlist is cleaned up, and an equivalence check with the original netlist is performed.

9.4.3.1 Step 1: Preparation

The first step contains all parts of the FSYN flow that are executed by the coordinating process using our open-source back end. This includes the first synthesis to a netlist of standard gates, the distribution of tasks via an HTTP API, the rewriting of the netlist with the following resynthesis, and the final equivalence check of the original and the optimized netlists.

First Synthesis In this first step of the FSYN flow, the design specified in the configuration file is synthesized to a standard gate-level library. FSYN can be applied to RTL level designs and to netlists. However, since the format of the netlist is important for the following phases of FSYN, synthesis has to be performed even if the design is already a netlist to get the expected output format. The TCL script controlling the Yosys synthesis is shown in Listing 9.4. The top-level module is set via the environment variable *PS_TOP* which is set by FSYN before the script is started. Most of the synthesis commands follow a basic Yosys synthesis. The call to *ABC* [6] is essential, as it maps the logic elements to a small set of gates which is built into Yosys: In fact, this guarantees that all gates are mapped to standard Verilog unary and binary logic operators (i.e., &, |, ^, and ~). For each of these basic operators, FSYN contains a set of properties that are used to check if a gate is redundant. The Verilog implementations of the properties introduced in Sect. 9.2.2 are presented in detail in Listing 9.5.

```
yosys read_verilog -sv -formal -DSYNTHESIS=1 work/step1.v
yosys hierarchy -check -top $::env(PS_TOP)
yosys flatten
yosys proc
yosys opt
yosys fsm
yosys opt
yosys memory
yosys opt
yosys techmap
yosys opt
yosys abc -g gates
yosys opt -full -share_all
yosys clean -purge
yosys write_verilog work/synthesized.v
```

Listing 9.4 TCL script for the first synthesis stage of FSYN

Task Server and Worker Assignment After the first synthesis is done, and the problem was split into tasks, the tasks have to be distributed to the workers. For this, FSYN starts an HTTP server that provides thread-safe access to a queue of all remaining tasks. The server provides an API with four endpoints that are used to synchronize and exchange tasks between the workers and the FSYN coordinating process. Each endpoint either returns a result in the form of JSON data or it can be used to POST JSON data to update the state of a task. The four endpoints of the API are:

/task/get	Retrieve an unsolved task from the queue.
/task/done	Mark a task as solved.
/done	Check if all tasks have been solved.
/remaining	Return the number of remaining tasks.

This API is started by the open-source back end after the design was synthesized and split into tasks. On startup, the API server reads the list of all tasks and inserts them into a thread-safe queue data structure. It also creates a hash map in which the state of all tasks (i.e., if the task was already solved) is kept. This hash map is initialized with all tasks set to the state of "not-solved."

After the task server is started, the number of tasks is compared to the number of available threads on all workers. FSYN now assigns each available worker a number of threads on which runners should be started. This assignment tries to fit as many tasks as possible to the workers in the *worker.json* file, starting with the first specified worker. Therefore, if the number of tasks is smaller than the number of all available threads, not every worker is assigned to start runners. All assigned tasks are stored in the *.assigned_worker.json* file. Following the task assignment, the coordinating process connects to every worker with assigned tasks to start the *psyn_client* application, which in turn spins off as many runner processes as have been assigned to this worker.

Now, the main phase of the FSYN flow starts in which all tasks are solved by the workers following the process presented in the description of the second phase. The API server is bundled with the FSYN application, and no additional software has to be installed. In particular, executing one *psyn run config* command on one of the workers is enough to start the full flow; everything else including synthesis, task dispatching, and PC is done automatically. While all tasks are being processed, the coordinating process waits until all runner processes on all workers are stopped. If all runner processes stopped, the final work is continued as presented in the description of the third phase.

9.4.3.2 Step 2: Distributed Execution

This section presents the parts of FSYN that are executed on the workers by the *psyn_client* application. The client application first reads the number of assigned threads from the *.assigned_worker.json* file and starts as many runners as were assigned to this worker.

Enriching Now, each runner process starts asking the API for an unsolved task by requesting the API endpoint */task/get*. If the API returns a task, the gates of this task must be annotated with properties. The SystemVerilog *bind* statement is not available in Yosys without commercial add-ons; therefore, we had to implement a different method to add the properties. Since every design must go through the first synthesis step, we can expect the netlist to always follow the same format. In particular, each gate of the netlist is implemented as a statement following the `assign _n_ = a OP b;` format, where *N* is an integer number corresponding to the gate numbers from the task and *OP* is one of the basic Verilog operands &, |, or ^. Additionally, gates can also come inverted in a format like `assign _n_ = ~(a OP b);`. We exploit this regular format by parsing each assign statement and, if the assign belongs to a gate from the task, keep a list of all gates and the type of this gate. After all gates are parsed, the properties are injected into the netlist by adding a module containing the properties for every required gate. The type of the module is injected depending on the type of the gate. By wiring the correct input and output signals into the module, the properties get bound to a gate. This property injection is implemented using the AWK text-processing language, and the result is a rewritten netlist containing the injected properties. We call this the *enriched* netlist which is saved into the *enriched.v* file in a task-specific local working directory.

Gate Properties The gate properties we use are based on the ones in [4] and are listed in Listing 9.5. We support the gate types *and*, *or*, *not*, and *xor*. For all gate types with two inputs, we check four basic assertions. A gate can be simplified if either the output is constant, which we denoted by the *_Y_0_p1* and *_Y_1_p2* assertions, or if it always follows one of the inputs, denoted by the *_Y_A_p3* and *_Y_B_p4* assertions. For the *not* gate, only the constant output assertions apply.

The properties follow the simple naming scheme *name_output_value_weight*. This naming scheme means that, if this property holds, the gate can be replaced by

```
1  `ifdef FORMAL
2  module and2_properties(input A, B, Y);
3      always @(*) begin
4          and_Y_0_p1: assert (Y == 1'b0);
5          and_Y_1_p2: assert (Y == 1'b1);
6          and_Y_A_p3: assert (!A || B);
7          and_Y_B_p4: assert (!B || A);
8      end
9  endmodule
10
11 module or2_properties(input A, B, Y);
12     always @(*) begin
13         or_Y_0_p1: assert (Y == 1'b0);
14         or_Y_1_p2: assert (Y == 1'b1);
15         or_Y_A_p3: assert (!B || A);
16         or_Y_B_p4: assert (!A || B);
17     end
18 endmodule
19
20 module not1_properties(input A, input Y);
21     always @(*) begin
22         not_Y_0_p1: assert (Y == 1'b0);
23         not_Y_1_p2: assert (Y == 1'b1);
24     end
25 endmodule
26
27 module xor2_properties(input A, B, Y);
28     always @(*) begin
29         xor_Y_0_p1: assert (Y == 1'b0);
30         xor_Y_1_p2: assert (Y == 1'b1);
31         xor_Y_A_p3: assert (!B);
32         xor_Y_B_p4: assert (!A);
33     end
34 endmodule
35 `endif
```

Listing 9.5 Verilog properties for the supported gate types

output. Additionally, each property is assigned a *weight*. This is required, as each gate is assigned multiple properties, of which some can hold at the same time. In this system, low weights mean a higher optimization potential. We assigned the weights using a simple method: Replacing a gate by a constant value (e.g., Line 4) is better than replacing it by a wire (e.g., Line 6).

PC and Redundant Gate Detection After the enriched netlist is written, the runner calls SymbiYosys to perform PC. Once sby is done with PC, it creates a log file with the PC results. This log file contains lines indicating if a given property failed during solving or if it holds. This means that we have to identify all properties for which no failing line exists in the log file. Like the property injection step before, we implemented this behavior with AWK by keeping track of a mapping of all properties and if they failed. Initially, all properties are set to holding. Then, all lines of the log file are analyzed if they contain information about a failing property. If all lines are processed, we eliminate all failing properties from the mapping and write out a list of all holding properties together with their gate into a file in the shared directory. However, as every gate has multiple properties that can hold at the same time, we also have to filter out the best holding property for each gate. This is done using the property weights described previously.

At this point, the runner is finished with the current task. It then posts the task to the */task/done* endpoint of the task API to mark this task as completed. Now, the runner checks if any tasks are left to do by requesting the *task/get* endpoint of the task API. If a task is returned by this request, the runner starts executing this task, else the runner stops itself.

9.4.3.3 Phase 3: Finalization

After all tasks are processed and all runners are stopped, the control is returned to the coordinating process. At this final stage, the only thing left to do is to rewrite the netlist with the optimization results, to resynthesize the rewritten netlist to perform additional optimizations and clean-up, and to perform the equivalence checking on the rewritten netlist.

Netlist Rewriting In this stage, the netlist is rewritten based on the files containing the holding properties for each gate which were produced by the runners. Similar as in the previous steps, the netlist rewriting is implemented using an AWK program that performs the rewriting directly in the Verilog netlist. First, this program parses the file containing all gates and their rules according to which the gate can be optimized. Next, it checks each line of the netlist for a gate that can be optimized. If such a gate is found, it is rewritten according to the rules presented in Sect. 9.2.2.

Resynthesis Rewriting the netlist results in two additional optimization cases. First, the rewritten netlist might contain new optimization potential for the traditional logic synthesis in Yosys. Second, the rewritten netlist might contain noise like chains of wire assigns (e.g., $a = b = c$) that can be reduced (e.g., to $a = c$). For this, we perform a second synthesis on the rewritten netlist using a very similar set of commands as in the first synthesis. The result of this second synthesis is also the final result of the FSYN flow.

Equivalence Checking Cutting out big parts of a design is an aggressive process that requires certain checks to ensure the correctness. To this end, our pipeline includes equivalence checking of the resulting netlist with the base one. Because the parts we cut out could only be removed under certain assumptions, we also have to annotate both netlists with the same assumptions, because we only expect them to be equivalent under these circumstances. Following the annotation, we can use the Yosys equivalence checking flow with a miter circuit to check for equivalence.

9.5 Experimental Results

In this section, we present experimental results demonstrating the optimization potential and performance characteristics of FSYN on various experiments ranging from arithmetic circuits all the way to RISC-V processors. First, the experimental setup including the compute cluster on which the experiments were conducted is presented in Sect. 9.5.1. Next, in Sect. 9.5.2, we find an advantageous number of gates that make up a task to decrease the overall runtime of FSYN. Then, in Sect. 9.5.3, we analyze the runtime behavior of FSYN by running selected designs from the EPFL synthesis benchmarks, each 20 times using randomized assumptions. Section 9.5.4 presents experiments reducing logic for a RISC-V processor. Furthermore, we compare the open-source against the commercial back end in Sect. 9.5.5. Next, we evaluate FSYNs rule filtering phase in Sect. 9.5.6, and finally, we evaluate the EC based back end in Sect. 9.5.7.

9.5.1 Experimental Setup

All experiments are conducted on a cluster of five compute servers (workers) running Ubuntu 20.04 LTS, connected over a local network and a shared NFS file system. Four of these workers have Intel Core i7-10700 8-Core processors with 16 threads, of which two have 64 GB RAM and two have 128 GB RAM. The last worker has an AMD EPYC 7713 64-Core Processor with 128 threads and 256 GB RAM.

9.5.2 Determining Favorable Task Sizes

As described before, after synthesizing the netlist, it is divided into multiple tasks, each representing a slice of all gates that have to be checked. Slicing the design into many smaller tasks has not only the advantage of enabling large-scale parallelism, but it also significantly speeds up the time required to solve each task. For example, solving all properties on all gates at once quickly overwhelms the solver, on the other hand, solving each gate individually is fast, but it also incurs some additional cost as the solver has to be started more often. Finding the sweet spot between solver speed and reducing the overhead of additional startups are therefore mandatory.

In this section, we experimentally find a favorable task size by applying FSYN on selected designs from the EPFL benchmarks with different task sizes. The EPFL benchmarks contain arithmetic and non-arithmetic designs including an adder, memory controller, arbiter, and router. Figure 9.6 shows a chart of the runtimes for the four largest designs from the *random_control* directory of the EPFL benchmarks, ordered from smallest (i2c) to largest (mem_ctrl). The runtimes are

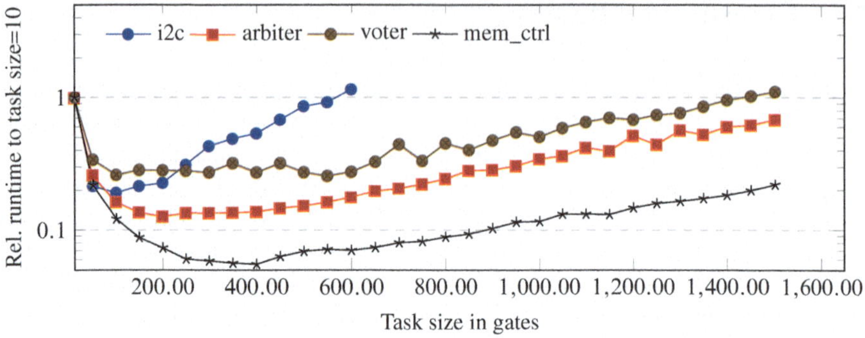

Fig. 9.6 Runtime for increasing task sizes

normalized, with the runtime of a task size of 10 serving as the baseline, and are shown on the y-axis. For each design, the task size is increased in steps of 50 starting with a task size of 10 all the way up to a task size of 1500. This process is stopped prematurely if either the task size is larger than the number of gates or the runtime surpasses the baseline value. The graph indicates that a small task size significantly increases the runtime and that runtimes sharply decrease with growing task sizes. However, this effect peaks at some point and starts to reverse for growing task sizes. A good task size depends on the size of the design, with larger designs requiring a larger task size. For example, the advantageous task size for the *i2c* design is at around 100 gates, while the advantageous task size for *mem_ctrl* is at around 400 gates.

9.5.3 Gate Reductions and Runtime

In this section, we apply FSYN on the EPFL benchmarks using randomized assumptions to analyze the reduction and runtime behavior for varying assumptions. For each selected design of the EPFL benchmarks, we generate 20 test cases with random assumptions on the top-level signals and observe the number of reduced gates and the runtime. By generating randomized assertions, we try to verify that the assumptions have little impact on the runtime of our FSYN flow. The results for the randomized runs are shown in Table 9.3. The first row lists the names of the selected benchmarks, and the second row shows the number of gates in this benchmark after the first synthesis step. The next rows list the selected task size, the summed runtime of all 20 runs, followed by the mean gate reductions and mean runtimes. Since it takes some time to start up the API server, a 3-second wait time is included in the original measurements. This waiting time distorts the following results for smaller designs, and therefore we subtracted these 3 seconds from all runtimes. The last row shows the average runtime per gate. The results in Table 9.3 show that FSYN can handle even large netlists with more than 40,000 gates in a

Table 9.3 Results of 20 runs for each selected EPFL benchmark with randomized assumptions

Benchmark	Ctrl	int2float	Router	Dec	Cavlc	Priority	i2c	Voter	Arbiter	mem_ctrl
Gates	99	201	207	304	633	641	1086	6164	11843	40876
Tasksize	80	80	80	80	80	80	80	200	200	400
Summed runtimes [s]	39	42	65	51	70	140	141	2716	2058	7765
Mean reduction [%]	83.44	65.97	93.57	81.65	76.12	18.73	7.46	1.44	1.69	4.44
Mean runtimes [s]	1.95	2.10	3.25	2.55	3.50	7.00	7.05	135.80	102.90	388.25
Time per Gate [s]	0.02	0.01	0.02	0.01	0.01	0.01	0.01	0.02	0.01	0.01

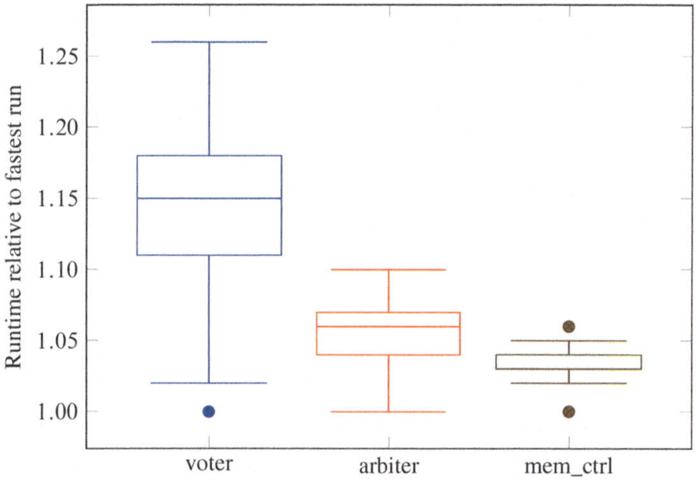

Fig. 9.7 Distribution of property checking runtimes for the 20 randomized runs

moderate amount of time. In particular, FSYN scales for these larger netlists with no drop in the time each gate takes to check.

In addition to the time each gate takes to check, we investigated how much the runtime varies for different assumptions. This is a very interesting metric, since knowing the runtime scales approximately with the size of the design and independently of the assumptions makes running FSYN more predictable. The runtimes of the three largest EPFL benchmarks are plotted in a box plot in Fig. 9.7. Note that the runtimes are plotted relative to the shortest measured runtime. In our experiments, we observed that the runtimes vary by at most 26% and that larger netlists result in less varying runtimes.

9.5.4 Reducing Netlists of RISC-V Processors

We used our FSYN flow to synthesize different configurations of the Ibex processor and compare the resulting gate counts. These configurations are based on the ones used in [4]. The first configuration, *Full*, puts the Ibex core without any additional assumptions into our tool and is used as a comparison value. *RV32im* limits the processor to instructions in the RV32im standard. *RV32i* removes the multiplication instructions. *Reduced Addressing* further removes any R-type instructions from the core. *Safety Critical* also removes JALR, AUIPC, FENCE, ECALL, and EBREAK. Lastly *RISC16* limits the core to the ADD, ADDI, AND, XOR, LUI, LW, SW, BEQZ, and JALR instructions. The results of applying FSYN to the IBEX core (c.f., Fig. 9.8) follow those in [4]. In particular, we see a similar large decrease in gate count when limiting the core to a set of instructions and once again having a noticeable decrease when further limiting the core to the RISC-16 ISA. All six configurations can be processed by FSYN in roughly the same time, taking around 8.5 minutes (approximately 500 seconds) for the PC using a task size of 80 and all machines described in Sect. 9.5.1.

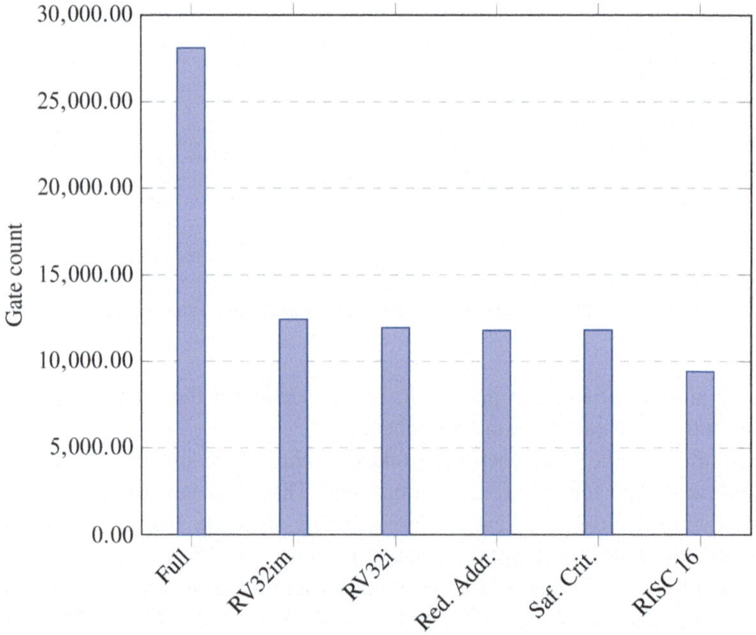

Fig. 9.8 Results for different RV32I subsets of the Ibex processor

9.5.5 Comparison to Commercial Tools

In addition to the open-source back ends, we implemented an FSYN back end using a well-known commercial formal verification tool. The general flow stays the same, but instead of injecting assertions in the netlist file, we can use SystemVerilog's full capabilities to bind the assertions to the gates. Similar to the open-source back end, the commercial back end can also solve multiple tasks in parallel. However, the number of concurrent processes is limited by the number of available licenses, which is 10 in our case.

The runtimes using the Yosys and the commercial back end for selected designs of the EPFL benchmarks are shown in Table 9.4. We applied FSYN to 5 designs of the EPFL benchmarks selected to cover a range of netlist sizes. Each design was optimized with FSYN, using a task size of 80. First, we ran FSYN using the open-source back end on the largest machine once using all the available machines from Sect. 9.5.1 (Row "Open-Source Full") and once with only 10 parallel tasks (Row "Open-Source 10"). Then, we ran FSYN twice using the commercial back end. In the first commercial run, we employed the tool's parallelization feature, utilizing all 10 licenses in parallel for solving (Row "Commercial Multi"). In this run, we split the assertions into subtasks using FSYN with a task size of 200. FSYN then passes these subtasks to the commercial tool, which distributes the assertions to parallel tasks. In the second run (Row "Commercial FSYN"), we applied the same settings of FSYN to split the assertions into smaller tasks which are then distributed to 10 workers without using the parallelization option of the commercial tool.

Both back ends result in the same removed gates. However, the open-source back end is significantly faster even when restricted to only 10 workers. We attribute the better performance of the open-source back end primarily to the behavior of the assertion checking process: The open-source back end can find multiple failing assertions at the same time, while the commercial tool checks all assertions individually. In a design considering realistic assumptions, most gates cannot be optimized, which means that most of the assertions are expected to fail. This particular setup, coupled with the requirement to verify a vast number of assertions which are most likely to fail, appears to result in a significantly reduced overhead in our case. However, from the data, we see that the commercial back end starts catching up with the open-source back end with increasing design sizes. Therefore, we believe that the commercial tool is geared toward solving "few" very complex assertions on large designs and not toward numerous very simple assertions. Considering this, we assume that the commercial back end starts outperforming the open-source back end for very complicated designs. However, the open-source back end will always have the advantage of practically unlimited scaling, since it is not limited by the number of available licenses. Moreover, exploring new task-finding heuristics and innovative formal problem formulations that make use of specific proof engines can only be achieved when the source code can be accessed.

Finally, we also compared the multi-threaded mode of the commercial tool (Row "Commercial Multi") to solving multiple tasks of the split-up problem using FSYN

Table 9.4 Runtime comparison between the open-source back end and the commercial back end

Benchmark	Ctrl	int2float	Cavlc	Adder	Arbiter
Gates	99	201	633	772	11843
Open-source full [s]	4	4	5	8	64
Open-source 10 [s]	4	4	5	8	234
Commercial multi 10 [s]	12	25	63	82	2677
Commercial FSYN 10 [s]	70	109	101	103	818

Table 9.5 Rule filtering results

Benchmark	Arbiter	mem_ctrl	32 × 32 Mul
Rules	47,356	161,844	27,216
Runtime PC 0 [s]	102	1,017	85
After filtering 40	1,339	49,859	12,038
Filtering time 40 [s]	21	97	22
Runtime PC 40 [s]	14	346	41
Runtime sum 40 [s]	35	443	63

at the same time (Row “Commercial FSYN”). We found that the commercial tools multi-threading offers less of a speed-up than solving more of the small subtasks if confronted with larger netlists. In this case, it pays off to let FSYN handle all parallelism.

9.5.6 *Rule Filtering*

In this section, we evaluate the new rule filtering phase of FSYN in the PC flow. All experiments were conducted using an AMD EPYC 7713 64-Core Processor with 128 threads and 256 GB RAM.

Table 9.5 shows experimental results for runs of FSYN with and without the new rule filtering. The first row shows the benchmark on which we applied FSYN. The first two columns are the arbiter and mem_ctrl benchmarks from the EPFL benchmark suite [1]. We randomly generated external constraints for them. The last column is an unsigned 32 × 32 bit multiplier (stages: simple partial product generator, array, Carry look-ahead adder) generated by [14]. As the external constraint for the multiplier, we set the lower 4 bit of the first input to 0.

The second row shows the number of all rules that have to be checked without the rule filtering. The third row shows the time it takes to check all rules without the rule filtering.

Next, in the fourth row, the number of rules which remain after the filtering (with 40 generated stimuli) is shown. The time the rule filtering took is shown in the fifth row, and the runtime to check the remaining rules is shown in the sixth row. Finally, the combined runtime of the rule filtering and the checking of remaining rules are reported in the last row.

Table 9.6 EC analysis for 32 × 32 multiplier

Gate	Result	PC runtime [s]	EC runtime [s]
N13330	Abort	>10,800	8
N13063	Holds	871	7
N13125	Holds	1,805	7
N13228	Abort	>10,800	7

Enabling rule filtering significantly reduces the number of rules that need to be formally proven, resulting in moderate (multiplier) or major (arbiter, mem_ctrl) runtime improvements.

9.5.7 FSYN with EC

In this section, we describe first experiments when utilizing the EC back end of FSYN. As a benchmark, we consider the same 32 × 32 bit multiplier as in the previous section with the identical external constraints (set lower 4 bit of the first input to 0). We analyzed the runtimes of the different PC problems in a configuration where we considered single PC problems and configured the time-out for solving a PC instance to 10 minutes (i.e., after that time the respective rule will be considered as failing, so we assume no optimization of the gate in terms of the rule can be performed). This time-out occurred for 12 out of 27,216 PC instances. Table 9.6 shows runtime data for some of these instances. The first column gives the line number of the gate inside the netlist. The second column reports the checking results for the rule. The third column shows the runtime of the PC problem, while the last column lists the runtime when employing EC. As can be seen, there are PC instances which still can be solved by increasing the time-out. However, there is even an instance which did not finish in more than 3 hours. Please note that all of these problems can be solved using EC within 7–8 seconds. Here, a more profound analysis is necessary.

9.6 Summary

In this chapter, we presented FSYN, a netlist optimization flow using formal methods. Given a set of external constraints (which can be viewed as external satisfiability DCs), our flow can eliminate gates from a netlist by formally checking if optimization rules can be applied to them.

We have formalized the optimization problem in terms of PC and EC to check optimization rules per gate. Depending on the validity of a gate's rule, it may either be removed as redundant, with its output connected directly to an input or a constant, or it may remain unchanged. Our implementation FSYN provides both, PC and

EC back ends, and leverages open-source tools for all steps, i.e., from synthesis to formal verification. Furthermore, we have demonstrated methods to accelerate the optimization process. This is achieved through parallelization of theoretically independent formal instances and a random-simulation-based approach that analyzes waveforms to refute rules.

Our experiments show that FSYN can significantly reduce netlists. Further, they show that FSYN efficiently utilizes the large parallelization potential of the netlist optimization problem. While FSYN is a promising step toward netlist optimization, additional research on formal gate-level optimization is necessary. This involves conducting a more comprehensive analysis of the relationship between task size and rule filtering. Furthermore, different EC strategies as well as the interplay of EC and PC need to be investigated. Finally, also other cost metrics and the potential in HW/SW optimization [24] are an interesting research direction.

References

1. L. Amarú, P.-E. Gaillardon, G. De Micheli, The EPFL combinational benchmark suite, in *International Workshop on Logic Synthesis* (2015)
2. K.A. Bartlett, R.K. Brayton, G.D. Hachtel, R.M. Jacoby, C.R. Morrison, R.L. Rudell, A.L. Sangiovanni-Vincentelli, A.R. Wang, Multi-level logic minimization using implicit don't cares. IEEE Trans. Comput. Aided Design Circ. Syst. **7**(6), 723–740 (1988)
3. A. Biere, A. Cimatti, E.M. Clarke, Y. Zhu, Symbolic model checking without BDDs, in *Tools and Algorithms for the Construction and Analysis of Systems* (Springer, Berlin, 1999), pp. 193–207
4. N. Bleier, J. Sartori, R. Kumar, Property-driven automatic generation of reduced-isa hardware, in *Design Automation Conference* (2021), pp. 349–354
5. D. Brand, Redundancy and don't cares in logic synthesis. IEEE Trans. Comput. **C-32**(10), 947–952 (1983)
6. R.K. Brayton, A. Mishchenko, ABC: An academic industrial-strength verification tool, in *Computer Aided Verification* (2010), pp. 24–40
7. R.K. Brayton, G.D. Hachtel, C. McMullen, A. Sangiovanni-Vincentelli, *Logic Minimization Algorithms for VLSI Synthesis* (Kluwer Academic Publishers, Dordrecht, 1984)
8. S. Chakraborty, K.S. Meel, M.Y. Vardi, A scalable and nearly uniform generator of SAT witnesses, in *Computer Aided Verification* (2013), pp. 608–623
9. L. De Moura, N. Bjørner, Z3: An efficient SMT solver, in *Tools and Algorithms for the Construction and Analysis of Systems* (Springer, Berlin, 2008), pp. 337–340
10. D. Große, L. Klemmer, D. Bonora, Using formal verification methods for optimization of circuits under external constraints, in *Design, Automation and Test in Europe* (2024), pp. 1–6
11. J.-H.R. Jiang, S. Devadas, Chapter 6 - logic synthesis in a nutshell, in ed. by L.-T. Wang, Y.-W. Chang, K.-T.T. Cheng, *Electronic Design Automation* (Morgan Kaufmann, Boston, 2009), pp. 299–404
12. L. Klemmer, D. Bonora, D. Große, Large-scale gatelevel optimization leveraging property checking, in *Design and Verification Conference and Exhibition Europe* (2023), pp. 86–93
13. S. Lee, H. Riener, G.D. Micheli, External don't cares in logic synthesis, in *International Workshop on Boolean Problems* (2022)
14. A. Mahzoon, D. Große, R. Drechsler, GenMul: Generating architecturally complex multipliers to challenge formal verification tools, in ed. by R. Drechsler, D. Große, *Recent Findings in Boolean Techniques* (Springer, Berlin, 2021), pp. 177–191

15. J. Marques-Silva, T. Glass, Combinational equivalence checking using boolean satisfiability and recursive learning, in *Design, Automation and Test in Europe* (1999), pp. 145–149
16. A. Mishchenko, R. Brayton, Simplification of non-deterministic multi-valued networks, in *International Conference on Computer-Aided Design* (2002), pp. 557–562
17. A. Mishchenko, R.K. Brayton, SAT-based complete don't-care computation for network optimization, in *Design, Automation and Test in Europe* (2005), pp. 412–417
18. P. Molitor, J. Mohnke, *Equivalence Checking of Digital Circuits: Fundamentals, Principles, Methods* (Springer Science & Business Media, Cham, 2007)
19. S. Muroga, Y. Kambayashi, H.C. Lai, J.N. Culliney, The transduction method-design of logic networks based on permissible functions. IEEE Trans. Comput. **38**(10), 1404–1424 (1989)
20. M.D. Nguyen, M. Thalmaier, M. Wedler, J. Bormann, D. Stoffel, W. Kunz, Unbounded protocol compliance verification using interval property checking with invariants. IEEE Trans. Comput. Aided Design Circ. Syst. **27**(11), 2068–2082 (2008)
21. H. Savoj, R.K. Brayton, The use of observability and external don't cares for the simplification of multi-level networks, in *Design Automation Conference* (1990), pp. 297–301
22. D. Shah, E. Hung, C. Wolf, S. Bazanski, D. Gisselquist, M. Milanovic, Yosys+nextpnr: An open source framework from Verilog to bitstream for commercial FPGAs, in *2019 IEEE 27th Annual International Symposium on Field-Programmable Custom Computing Machines (FCCM)* (2019), pp. 1–4
23. M. Sheeran, S. Singh, G. Stålmarck, Checking safety properties using induction and a SAT-solver, in *International Conference on Formal Methods in CAD* (2000), pp. 108–125
24. S.G. Sørensen, C. Bartsch, D. Stoffel, W. Kunz, Generation of formal CPU profiles for embedded systems, in *VLSI of System-on-Chip* (2022), pp. 1–6
25. C. Wolf, Yosys open synthesis suite. https://yosyshq.net/yosys/

Chapter 10
Conclusion and Outlook

After decades of processor performance gains, as predicted by Gordon Moore, improvements in the manufacturing process of integrating circuits are starting to slow down. However, the demand for and expectations from new processors are still growing rapidly. In an attempt to cope with the end of the Moore era, the industry is pivoting toward highly application-specific designs. This development, together with rising design sizes and complexities, creates major challenges for the Electronic Design Automation (EDA) industry. From the design side, much work has been done recently to keep up with these developments. This includes new and much more productive Hardware Description Languages (HDLs) and Virtual Prototypes (VPs), which allow starting software development sooner. Yet, considering the complete hardware development process, about 50% of the work lies not in the design phase but in the verification and debugging phases.

This book presents several versatile hardware analysis techniques covering some of the most time-consuming phases of the design process. In particular, it proposes new techniques in the following hardware development phases, which will be summarized below:

1. Verification
2. Debug
3. Post-synthesis Optimization

Several contributions targeting the verification phase are presented. The first contribution in the verification phase is a novel processor verification technique based on equivalence checking. By generating two distinct but equivalent programs with the help of a formal model, bugs inside processor designs can be found by comparing the result of the two programs. In addition, a framework for formally verifying processor microcode is presented. Using this framework, a microcoded RISC-V platform was successfully verified. The microcode for this platform, which is based on the One Instruction Set Computer (OISC) principle, was developed using a VP. This VP significantly reduces the development time and effort and

L. Klemmer, D. Große, *Versatile Hardware Analysis Techniques*,
https://doi.org/10.1007/978-3-031-83093-8_10

allows efficient design space exploration. The next contribution is the Waveform Analysis Language (WAL), a Domain Specific Language (DSL) for analyzing waveforms. WAL provides many language features that are not commonly found in traditional programming languages and are specifically targeted toward waveform analysis. In particular, these features include information about timing and design hierarchy, two concepts that are very specific to hardware design. The practicability and effectiveness of WAL are demonstrated in this book in several case studies. Those case studies showcase how WAL can be used for typical functional and extra-functional verification problems. Examples of such verification problems are processor and cache performance analysis, as well as processor pipeline analysis. Further, we present how WAL can be integrated into the design phase by embedding it into an HDL using analysis passes.

Next, for the debugging phase, this book presents a novel debugging flow based on WAL. This debugging flow utilizes WAL's ability to inject new signals into already existing waveforms. The values of these signals are calculated by running WAL programs which drive the signals. With these virtual signals, an entirely new debugging flow becomes possible, in which users can identify a bug, fix it directly within the verification environment, and finally verify the bug fix inside the same environment. This highly interactive debugging flow can significantly reduce debugging times, since it allows hardware engineers to stay and work within the same environment without having to go all the way through the design, simulation, and verification steps.

Finally, for the post-synthesis optimization phase, this book proposes FSYN, a netlist optimization approach based on external constraints. User-supplied external constraints unlock optimization potential in netlists. For example, certain gates of a processor can be unnecessary if a specific instruction will never be a part of the executed programs. In this case, these unnecessary gates can be safely removed from the netlist, thus leading to a potentially significant size reduction. However, finding these gates is challenging due to the size of netlists and the virtually infinite number of input combinations. The FSYN flow utilizes formal property and equivalence checking to reliably and efficiently find unnecessary gates. The efficiency and effectivity of FSYN are demonstrated in several case studies, showing significant netlist size reductions and high optimization potential in regards to the optimization runtimes.

All presented techniques are implemented and evaluated in several experiments and case studies. Further, all of the presented techniques are available as open-source software to stimulate further research and to enable reproducibility.

In summary, the contributions in this book significantly enhance the current state-of-the-art of hardware analysis. By covering a wide range of hardware analysis problems, this book contributes to reducing the impact of current bottlenecks in the hardware development process, allowing further performance improvements in the post-Moore era.

Outlook The results presented in this book clearly show the effectiveness and benefits of the proposed techniques. However, most of these techniques are only initial work that opened the door to further research in terms of effectiveness and performance. In the following, we briefly discuss some ideas for future improvements to the presented work.

- Optimizing WAL core: Currently, the reference implementation of WAL is not targeted at high performance. Developing a high-performance WAL interpreter or compiler would unlock many additional analysis applications and would significantly increase WAL's applicability to industry scale problems. In particular, we see many potential performance improvements in new waveform formats and data structures that are designed with waveform analysis in mind.
- Improving the user experience for WAL: This in particular means providing WAL front ends that are more approachable for hardware designers and verification engineers. We present first steps in this direction with the proposed HDL integration and the SystemVerilog Assertions (SVA) compiler but more steps are necessary. In this direction, we plan to further develop the SVA compiler, possibly extending it to a larger SystemVerilog subset.
- Integrating WAL into more EDA tools: We have presented how WAL on its own can be used to perform powerful waveform analysis in several case studies. Nonetheless, we see much potential in integrating WAL into more (open-source) EDA tools. First steps in this direction have been taken with our HDL integration. However, more work is planned or already underway. One prominent example of this is a tight integration with waveform viewers such as Surfer. Such a tight integration could provide users with a rich and interactive analysis and debugging workflow.
- Utilizing the formal microcode verification framework for microcode synthesis: One positive side effect of using Satisfiability Modulo Theories (SMT) solvers for formal verification is that it is relatively straightforward to extend the model to allow code synthesis. While synthesizing complex microcode routines is very challenging, our microcode verification framework provides a suitable starting point to further explore this line of research. Preliminary experiments with synthesizing code using our formal models have shown promising results. Nevertheless, especially synthesizing deeply nested code and code containing loops is still challenging. While more work is needed, microcode synthesis allows, for example, quickly adapting to changing design constraints or supporting highly custom micro-architectures easily.
- Combining the strengths of Property Checking (PC) and Equivalence Checking (EC) based optimization of external constraints in the proposed FSYN optimization flow: While the PC flow allows efficiently checking a large number of optimization rules in parallel, we demonstrate that the EC flow can be faster for long-running problems. By combining the PC and EC flows, we could thus achieve a better overall performance, as long-running tasks can be handed over to the EC flow to potentially solve them in the given time budget.

Index

L. Klemmer, D. Große, *Versatile Hardware Analysis Techniques*,
https://doi.org/10.1007/978-3-031-83093-8

The manufacturer's authorised representative in the EU is Springer Nature Customer Service Centre GmbH, Europaplatz 3, 69115 Heidelberg, Germany. If you have any concerns regarding our products, please contact ProductSafety@springernature.com

Printed and bound by CPI Group (UK) Ltd, Croydon, CR0 4YY
01/10/2026
02225960-0001